T0180668

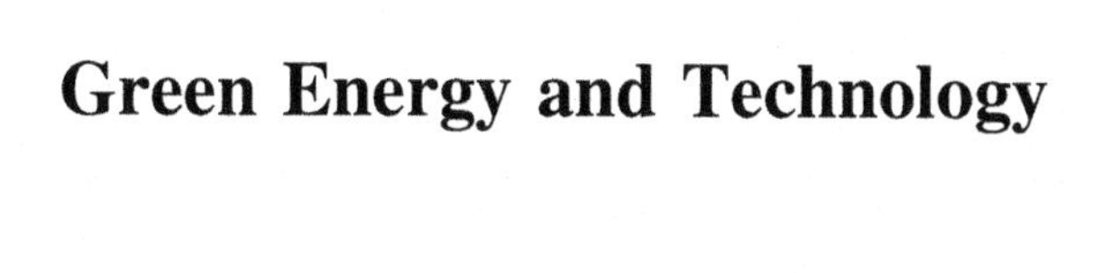

Green Energy and Technology

For further volumes:
http://www.springer.com/series/8059

Djamila Rekioua

Wind Power Electric Systems

Modeling, Simulation and Control

Djamila Rekioua
Universite de Bejaia
Bejaia
Algeria

ISBN 978-1-4471-7237-6 ISBN 978-1-4471-6425-8 (eBook)
DOI 10.1007/978-1-4471-6425-8
Springer London Heidelberg New York Dordrecht

Softcover reprint of the hardcover 1st edition 2014

Printed on acid-free paper

Springer is part of Springer Science+Business Media (www.springer.com)

Preface

Generally, a book cannot be written without assistance. So, therefore, I must thank all the many people who helped me. I would particularly extend special thanks to my colleagues Prof. A. M. Tounzi, T. Rekioua, and Dr. K. Idjdarene for cooperation in common projects in renewable energy, and especially in wind energy. In addition, thanks are due to Springer-Verlag for publishing this book.

Bejaia, Algeria Djamila Rekioua

Contents

Acronyms

General

v, **V**	Voltage in instantaneous and vector notation
i, **I**	Current in instantaneous and vector notation

Subscripts

d, q	Quantities in d-axis and q-axis
α, β	Quantities in α-axis and β-axis
s, r	Stator and rotor
a, b, c	Quantities in phases a, b, and c

Superscripts

AC	Alternative Current
AFLC	Adaptive Fuzzy Logic Controller
ANFIS	Adaptive Neuro-Fuzzy Inference System
ANN	Artificial Neural Networks
BDFIG	Brushless Doubly Fed Induction Generator
BDFM	Brushless Doubly Fed Machine
BN	Big Negative
BP	Big Positive
CAES	Compressed Air Energy Storage
CIEMAT	Centro De Investigaciones Energéticas, Medioambientales Y Tecnológicas
CPG	Claw-Pole Generator
DC	Direct Current
DFIG	Doubly Fed Induction Generator
DFIM	Doubly Fed Induction Machine
DOD	Depth of Discharge
DPC	Direct Power Control
DPCWS	Direct Power Control of Wind System

DSIM	Dual Stator Induction Machine
DSWIM	Dual Stator Winding Induction Machine
DTC	Direct Torque Control
EG	Electrical Generator
ELC	Electronic Load Controllers
EleS	Electrical Energy Storage
Elmes	Electromagnetic Energy Storage
ES	Energy Storage
FBS	Flow Batteries Storage
FES	Flywheel Energy Storage
FLA	Flooded Lead Acid
FLC	Fuzzy Logic Controller
FOC	Field Oriented Control
FSWT	Fixed Speed Wind Turbine
FT	Factor Temperature
G	Gearbox
GTO	Turn-Off Thyristor
H_2	Hydrogen
HAWT	Horizontal-Axis Wind Turbine
HCS	Hill-Climb Searching
HES	Hydrogen Energy Storage
HPS	Hybrid Power Systems
ICS	Indirect Control Speed
IGBT	Insulated Gate Bipolar Transistor
LAS	Lead-Acid Storage
LIG	Linear Induction Generator
Li-Ion	Lithium Ion
LIS	Lithium Ion Storage
MHDTC	Modulated Hysteresis Direct Torque Control
MN	Means Negative
MP	Means Positive
MPPT	Maximum Power Point Tracking
MS	Mechanical Storage
Nas	Sodium-Sulfur
Nass	Sodium-Sulfur Storage
NCS	Nickel–Cadmium Storage
Nicd	Nickel–Cadmium
Ni–Zn	Nickel–Zinc
NN	Neural Network
O_2	Oxygen
OTC	Optimal Torque Control
P&O	Perturb & Observe
PCU	Power Control Unit
PE	Power Electronic
PHES	Pumped Hydro Energy Storage

PI	Proportional Integral
PMSG	Permanent Magnet Synchronous Generator
PMTFM	Permanent Magnet Transverse-Flux Machines
PSBS	Polysulphide Bromide Storage
PSF	Power Signal Feedback
PWM	Pulse Width Modulation
RBFN	Radial Basis Function Network
SAWS	Stand-Alone Wind System
SC	Scalar Control
SCIG	Squirrel Cage Induction Generator
SCWS	Scalar Control of Wind System
SES	Super Capacitor Energy Storage
SG	Synchronous Generator
SIV	Speed Indirect Control
SMC	Sliding Mode Control
SMES	Superconducting Magnetic Energy Storage
SN	Small Negative
SOC	State of Charge
SP	Small Positive
SRG	Switched Reluctance Generator
STATCOM	Static Synchronous Condenser
TC	Torque Coefficient
TCR	Thyristor Controlled Reactor
TES	Thermal Energy Storage
TSC	Thyristor Switched Capacitor
Tsos	Transmission System Operators
TSR	Tip Speed Ratio
VAWT	Vertical-Axis Wind Turbine
VC	Vector Control
VCWS	Vector Control of Wind System
VR	Vanadium Redox
VRB	Vanadium-Redox Flow Battery
VRLA	Valve-Regulated Lead Acid
VSI	Voltage Source Inverter
VSWT	Variable Speed Wind Turbine
Z	Zero
ZEBRA	Zero Emission Battery Research Activity
Znbr	Zinc Bromine

Notations

A_{pv}	Size of photovoltaic generator
A_{wind}	Swept area
B_{ls}	Torsion coefficient on the low shaft
C	Scale factor
C_{DC}	DC-link capacitance
C_1 and C_2	Capacities of the battery at different discharge-rate states
C_{batt}	Battery capacity
C_{10}	Rated capacity
$C_T(\lambda, \beta)$	Torque coefficient
$C_P(\lambda, \beta)$	Power coefficient
DOD	Depth of discharge
$e(x)$	Difference between the controlled variable x and its reference $\hat{x}$
$E_{a,b,c}$	Induced f.e.m in the stator phase windings
E_b	Zero-current voltage of the battery charged
E_c	Kinetic energy stored in the flywheel
E_{c0}	Initial energy of the flywheel
E_{cref}	Reference energy for the flywheel
$\overline{E_L}$	Annual average energy
$\overline{E_{pv}}$	Annual average values of PV monthly contribution
E_T and E_Φ	Torque and flux errors
$\overline{E_{\text{wind}}}$	Annual average values of wind monthly contribution
$E_{\text{wind},m}$	Monthly energy produced by the system per unit area
$E_{\text{load},m}$	Monthly energy required by the load
E_{wind}	Energy produced by wind generator
ff_0	Calm frequency
f_n	Nominal stator frequency
f_{pv}	Fraction of load supplied by the photovoltaic energy
f_s	Stator frequency
E_{wind}	Fraction of load supplied by the wind energy
E_s	Solar radiation on tilted plane module
g	Slip
I	Angle of incidence
$I_{a,b,c}$	Current phase voltages

I_{batt}	Battery's current
I_{d0} and I_{q0}	Magnetising currents along the d and q axis
I_{dr} and I_{qr}	d-q rotor currents
I_{ds} and I_{qs}	(d, q) components of the stator current
I_g	Grid current
I_{load}	Load's current
I_s	Stator current
I_{10}	Current which corresponds to the operating speed
J_{FESS}	Inertia flywheel
J_{mot}	Inertia of the AC motor
J_t	Inertia of the turbine
$J_{t_{hs}}$	Inertia of turbine (the high shaft)
J_w	Moment of inertia of the flywheel
K	Constant
K_i	Polarization voltage
K_{opt}	Coefficient which depends on the ratio of tip speed and optimal power coefficient maximum
K_R	An experimental constant
K_t	Friction coefficient of the turbine
$K_{t_{hs}}$	External friction coefficient of the turbine
L	Stator phase inductance
L_c	Cyclic inductance
L_s	Stator leakage inductance
L_m	Magnetising inductance
M	Month of the year
N	Peukert constant
N_b	Battery number
N_{inp}	Number of inputs
N_j	Number of days of autonomy
N_{phase}	Number of phases
N_{pv}	Number of photovoltaic generator
N_{wind}	Number of wind generator
P	Number of pole pairs
P_{diesel}	Power delivered from the diesel generator
P_g	Power fixed at the grid
P_{jr}	Power fixed at the grid
P_{js}	Power fixed at the grid
P_{load}	Power required by the load
P_{pv}	Power delivered from the PV
P_{ref}	Active power control
P_s	Stator power
$P_{\text{st\_ref}}$	Reference value of the storage system output power
P_{mec}	Mechanical power
$P_{\text{tb-max}}$	the maximum power turbine

P_{unm}	Unmet load
P_{wind}	Power delivered from the wind turbine
P_{required}	Total required power
Q	Accumulated ampere-hours divided by full battery capacity
Q_{ref}	Reactive power control
R	Blade's radius
R_s	Stator resistance
R_r	Rotor resistance
R_{batt}	Internal (ohmic) resistance of the battery
R_0	Total internal resistance of a fully charged battery
R_{ins}	Inserted resistance
R_m	Core loss resistance (Ω)
S	Apparent short circuit power
SOC	State of charge
S_{wind}	Size of wind generator
T	Discharging time
$T_{a_{hs}}$	Aerodynamic torque of the generator reduced the high shaft
T_{em}	Average value of torque
T_{diesel}	Control signal of the relays from diesel
T_j	Cell temperature
T_g	Electromagnetic torque of the generator reduced to the low shaft
T_{ls}	Low speed shaft torque
T_{pv}	Control signal of the relays from PV energy
T_{wind}	Control signal of the relays from wind energy
T_{jref}	Reference cell
U_n	Nominal voltage
$V_{a,b,c}$	Machine phase voltages
V_{batt}	Battery voltage
V_{ds} and V_{ds}	(d, q) components of the stator voltage
V_N	Nominal stator voltage
V_{oc}	Open circuit voltage of a battery cell when fully charged
V_s	Stator voltage
$V_{s\alpha}$ and $V_{s\beta}$	(α, β) component of stator voltage
V_{wind}	Wind speed
$\tilde{v}_{\text{wind}}$	Fictive measure of the wind speed
X	Controlled variable
$\hat{x}$	Reference of x
X_c	Number of possible combinations
X_m	Magnetizing reactance
X_s	Stator leakage reactance
X_r	Equivalent rotor leakage reactance
α_{sc}	Temperature coefficient of short-current
β	Blades pitch angle
γ	Gamma function

θ_n	Digitized signals
Θ	Position of the rotor
λ_x	Positive constant
λ_{opt}	Optimal tip speed ratio
ρ	Air density
γ_ψ	Flux vector position
ε_ω	Tracking error
ΔT	Heating of the accumulator
Δh_p and Δh_q	Hysteresis band
ΔP and ΔQ	Active and reactive signals
ΔP_{diss}	Power dissipated in the dump load
η_r	Reference efficiency of the photovoltaic generator
η_{pc}	Power conditioning efficiency
η_{gen}	Photovoltaic generator efficiency
η_{coul}	Coulombian efficiency
ω_t	Turbine speed
ω_g	Generator speed
ω_r	Rotor speed
ω_s	Stator speed
ω_{ref}	Reference angular velocity
ω_{opt}	Optimal reference angular velocity
ω_{ls}	Low shaft speed
Ω_w	Speed of the flywheel
Ω_{ref}	Reference speed for the flywheel
Φ_{sd} and Φ_{sq}	(d, q) components of the stator flux
Φ_{rd}, Φ_{rd}	(d, q) rotor flux
σ	Leakage coefficient

Introduction

Djamila Rekioua

Wind systems comprise generally a wind turbine, which is installed on top of a tall tower, collects kinetic energy from the wind and converts it to electricity. The most important advantages of the wind systems are:

- They use clean and free energy
- They require no connection to an existing power source or fuel supply
- They could be combined with other power sources to increase system reliability
- They consume no fossil fuels
- They could be installed and upgraded as wind firm, more wind turbine could be added as power demand increases

The power price has considerably decreased since the last decade. This leads to a large-scale application of wind systems in several promising areas. Compared with conventional fossil energy sources, small wind energy systems are the best option for many isolated or rural areas applications around the world. However, because wind energy is an intermittent and a variable source of energy, stand-alone turbines generally can use another source of energy to provide constant power, such as solar photovoltaic or hydro.

Aims of the Book

Many books currently on the market are treating on the wind energy and wind energy conversion systems. This book treats not only elementary definitions on wind energy, but also optimization, modelization, simulation, and various linear and nonlinear controls applied to wind systems with applications under MATLAB/ Simulink.

The main objective of this book is to enable all students for graduation and postgraduations especially in the fields of electrical engineering to quickly understand the concepts of wind systems, provide models, control, and optimization. We present in first part, some stand-alone wind applications, such as rural electrification and pumping. And in second part, we give some applications in grid-connected system. Mathematical models are given for each system and a corresponding example under MATLAB/Simulink package is given at the end of each section. Various examples are given for an eventual implementation under DSPACE package. Some electrical machine control approaches, such as vector control, direct torque control, and fuzzy logic controllers are introduced in different drive systems. Furthermore, in order to optimize the wind operation, intelligent techniques are developed. By writing this book, we complete the existing knowledge in the field of wind and the reader will learn how to make the modeling and the optimization of the most used stand-alone and grid-connected wind applications by applying different control strategies.

How the Book is Organized?

The book is organized through seven chapters as follows:

- Chapter 1 is intended as an introduction to the subject. It defines the wind process, introduces the main meteorological elements, the wind velocity, and presents an overview of wind systems (stand-alone systems and grid-connected systems). This chapter includes also presizing and maintenance of wind systems.
- Chapter 2 focuses on wind energy conversion and power electronics modeling. The different structures of converters used in wind systems are presented. Some applications are given under Matlab/Simulink.
- In Chap. 3, a detailed review on the most used algorithms to track the maximum power point is presented. Several maximum power point tracking (MPPT) algorithms have been developed to track the maximum power point of the wind turbine. We present the most used in wind energy conversion. Some simple Matlab/Simulink examples are given.
- In Chap. 4, a description and modeling of the storage device is showed. The study describes the different storage used in wind energy. We present some energy storage applications under Matlab/Simulink.
- In Chap. 5, several nonlinear control of wind turbine systems are presented. The mechanical and the electrical part of wind turbine is detailed with application under Matlab/Simulink of some nonlinear control.
- The Chap. 6 is devoted to hybrid wind systems. The chapter describes the different configurations and the different combinations of hybrid wind systems. Different synoptic schemes and simulation applications are also presented.
- Finally, in Chap. 7, we present some wind examples with simulation and experimental tests.

Chapter 1
Conversion Wind System Overview

1.1 Introduction

Wind energy is a renewable, clean, and free energy source for energy production. Wind energy conversion system (WCES) requires no connection to an existing power source, and they could be combined with other power sources to increase system reliability and could be installed and upgraded as wind firm; more wind turbine could be added as power demand increases. The first new modern wind turbines were in 1979, and their power capacities were around 10–30 kW. Wind power technologies can be classified as follows:

- by axis wind turbine [horizontal axis wind turbines (HAWT) or vertical axis wind turbines (VAWT)]
- by localization (onshore or offshore).

1.2 Global Structure of a Conversion Wind System

The simplest structure of WCES is presented in Fig. 1.1. The aerogenerator and power electronics interface are the main part of the WECS. Wind turbine converts the mechanical energy to electrical one through electrical generator (EG), and power electronic converters control the aerogenerator. The rotor is made up of blades which convert wind energy into mechanical one. It turns the shaft at low speeds, so we use gearbox to increase speeds required by EGs. Some turbines as small-scale turbine do not require a gearbox, and they use a direct-drive system.

1.2.1 Wind Speeds

Wind speed, or wind velocity, is the fundamental output of a conversion wind system. It depends on metrological conditions. It is commonly measured with an

D. Rekioua, *Wind Power Electric Systems*, Green Energy and Technology,
DOI: 10.1007/978-1-4471-6425-8_1,

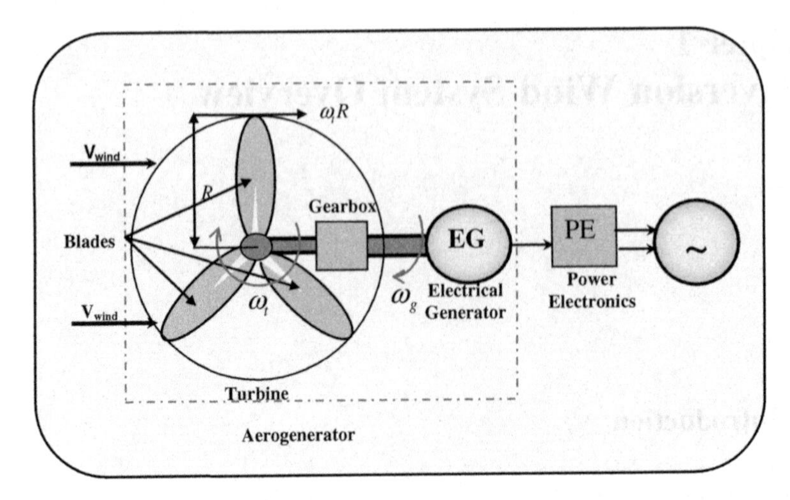

Fig. 1.1 Structure of a conversion wind system. R is the blade's radius, V_{wind} is the wind speed or air velocity, ω_t is the turbine speed, ω_g is the generator speed, G represents the gearbox, EG represents the electrical generator, and PE represents power electronics

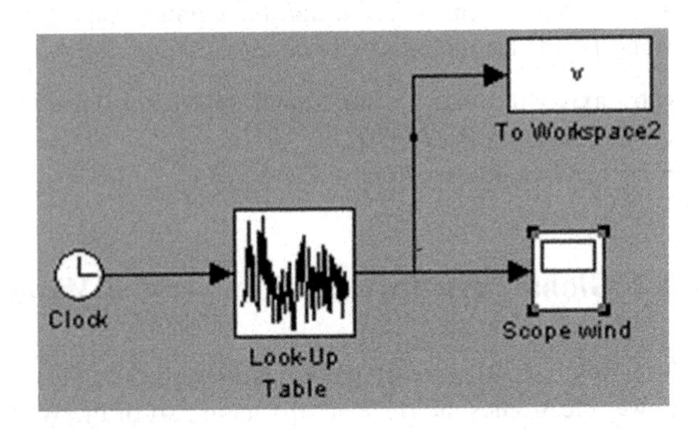

Fig. 1.2 Wind variations under MATLAB/Simulink

anemometer or by a weather station in (m/s) or (km/h). We modeled it under MATLAB/Simulink (Figs. 1.2 and 1.3).

We defined the following speeds (Fig. 1.4):

- Cut-in wind speed (around 3.5 m/s): At this speed, the turbine starts to rotate.
- Nominal wind speed or rated wind speed (between 11 and 17 m/s): At these speeds, the power output reaches the limit that the EG is capable of.
- Cut-out wind speed (between 17 and 30 m/s): As wind speeds increase, there is a risk of damage to rotor, and we are obliged to use a braking system to bring the rotor to a standstill.

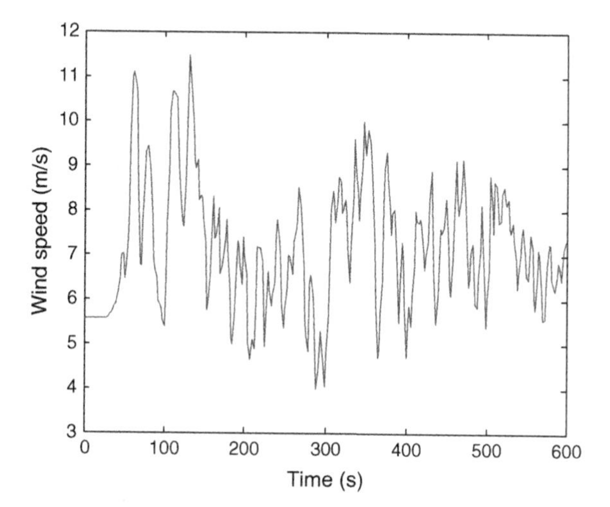

Fig. 1.3 Wind speed profile obtained under MATLAB/ Simulink

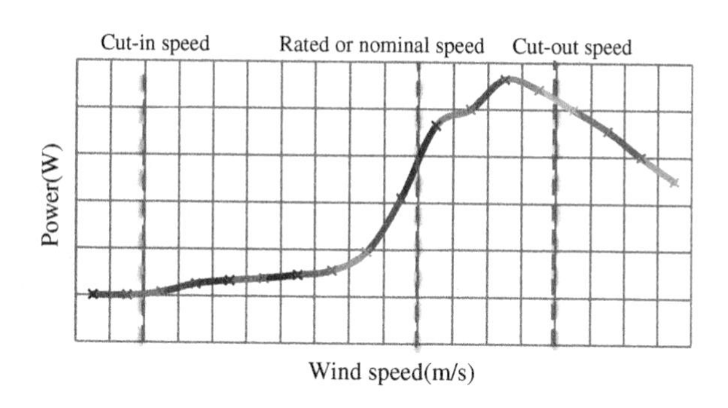

Fig. 1.4 Typical curve for a wind turbine

1.2.2 Aerogenerator

1.2.2.1 Wind Turbine Axis

Wind turbines can rotate about either HAWT or VAWT (Darius, Savonius).

Horizontal Axis Wind Turbines

HAWT is the most used. It is mounted on towers (Fig. 1.5). The main advantages of the HAWT are its high efficiency and low cost/power ratio. Its drawbacks are the complex design and the difficulties in maintenance because generator and gearbox should be mounted on a tower.

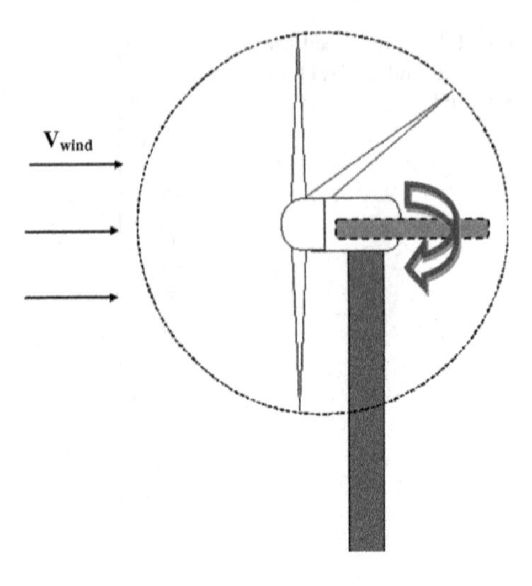

Fig. 1.5 Horizontal axis wind turbine description

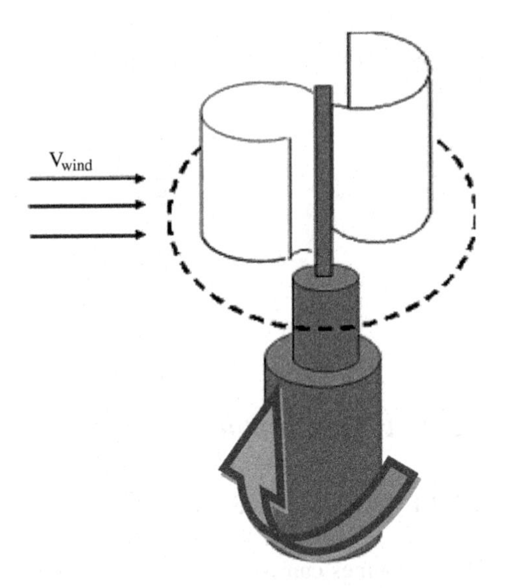

Fig. 1.6 Savonius vertical axis wind turbine description

Vertical Axis Wind Turbine

The first windmills built were based on vertical axis structure. There is Savonius rotor (Fig. 1.6) and Darius rotor (Fig. 1.7). In VAWT, the maintenance is easy and it receives wind from any direction. Its blade design is simple and has a fabrication of low cost. Its main drawbacks reside on requiring a generator to run in motor mode at start and having lower and oscillatory component in the aerodynamic torque.

We can resume the wind turbines classification which depends on orientation of the shaft and rotational axis in Table 1.1.

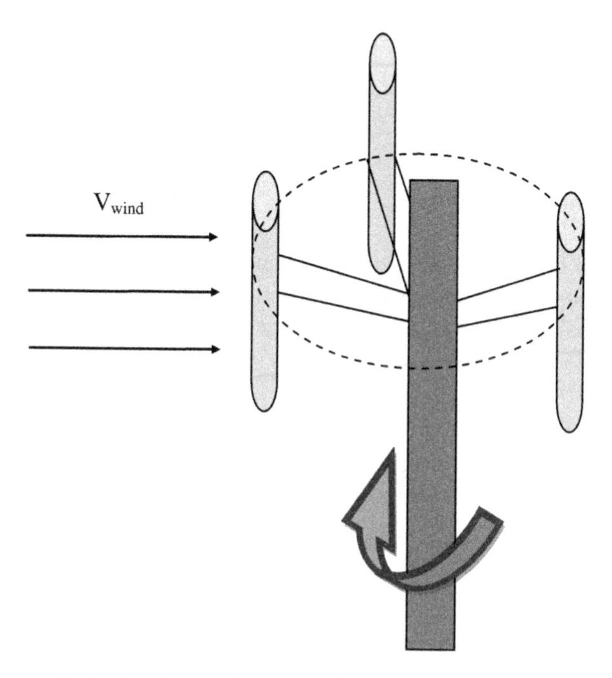

Fig. 1.7 Darius vertical axis wind turbine description

Table 1.1 Wind turbine classification

Wind turbines		Efficiency (%) [1]	Description
HAWT	One blade	13	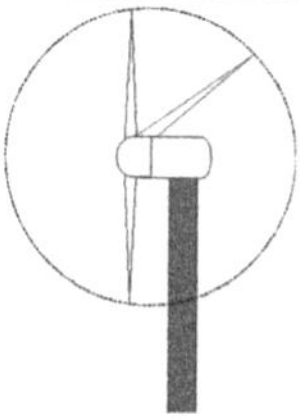
	Two blades	47	
	Three blades	50	
VAWT	Darius	40	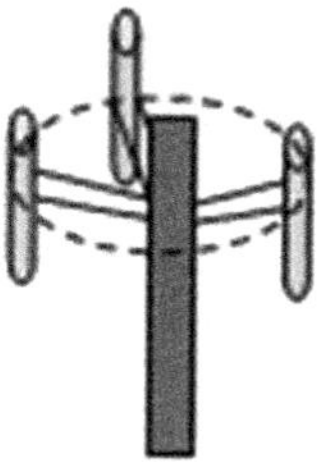
	Savonius	16	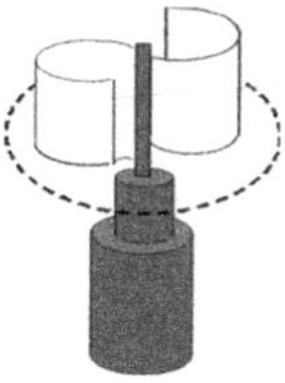

Table 1.2 Tip speed ratio number

λ	Value
1–2	Low
10	High

1.2.2.2 Mechanical Gearbox

The mechanical connection between an EG and the turbine rotor may be direct or through a gearbox. In fact, the gearbox (G) converts the turning speed of the blades ω_t to a rotational speed ω_g of an electrical generator (EG):

$$\omega_t = \frac{\omega_g}{G} \tag{1.1}$$

We defined the tip speed ratio (TSR) for wind turbines as ratio of the rotational speed of the tip of a blade $\omega_t R$ to the actual wind speed V_{wind} (Table 1.2).

$$\lambda = \frac{\omega_t \cdot R}{V_{\text{wind}}} \tag{1.2}$$

Modern horizontal axis wind turbine uses generally λ of nine to ten for two-bladed rotors and six to nine for three-bladed rotors [1].

1.2.2.3 Power Coefficient

We defined the power coefficient of a wind turbine as a measurement of how the wind turbine converts the energy in the wind into electricity. It depends on wind turbine axis (Fig. 1.8).

We can express power coefficient with different expressions [2, 3]:

Power Coefficient in Terms of Axial Induction Factor (a)

The fraction by which the axial component of velocity is reduced is known as the axial induction factor (a). If the free stream velocity is V_1 and the axial velocity at the rotor plane is V_2, then the axial induction factor is given as follows (Fig. 1.9):

$$a = \frac{V_1 - V_2}{V_1} \tag{1.3}$$

The power coefficient C_p can be expressed in terms of axial induction factor (a) as follows [2]:

$$C_p = 4a \cdot (1 - a)^2 \tag{1.4}$$

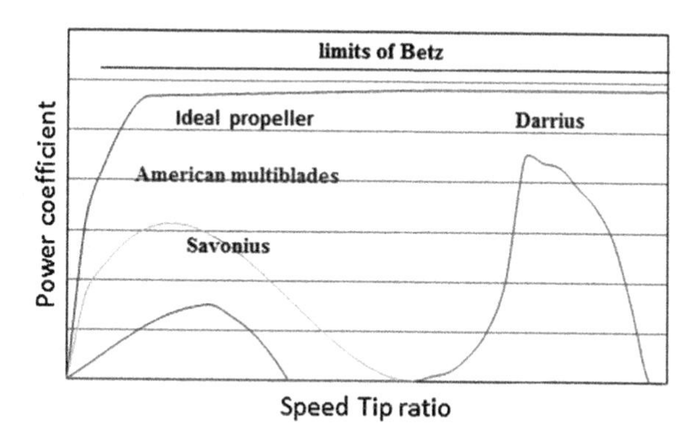

Fig. 1.8 Power coefficient for different wind turbine axis

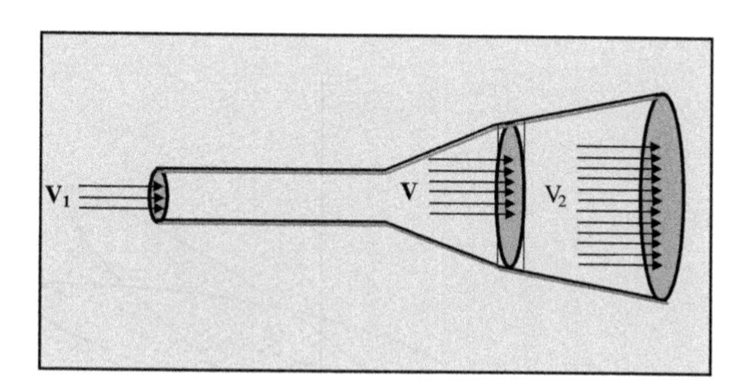

Fig. 1.9 Wind velocity direction through the rotor plane of wind turbine

Power Coefficient in Terms of the Tip Speed Ratio (TSR)

The power coefficient C_p can be expressed in terms of the TSR λ as follows [1, 2]:

$$C_p(\lambda) = -0.2121 \cdot \lambda^3 + 0.0856 \cdot \lambda^2 + 0.2539 \cdot \lambda \tag{1.5}$$

This expression is used in the case of Savonius turbine or small American multiblades (less than 01 kW). It is shown in Fig 1.10.

The Power Coefficient in Terms of TSR λ and Pitch Angle Beta β

The incidence angle is the angle between the wind relative velocity (a vector velocity resulted from the wind axial rotor velocity and the wind rotational velocity) and the rotation plane. We defined the pitch angle and the angle of incidence i shown in Fig. 1.11 as follows:

$$i = \operatorname{arct} g\left(\frac{1}{\lambda}\right) = \operatorname{arct} g\left(\frac{R \cdot \omega_{\text{tb}}}{v_{\text{wind}}}\right) \tag{1.6}$$

where V_{res} is a vector velocity resulted from the wind axial rotor velocity and V_{rot} is the wind rotational velocity.

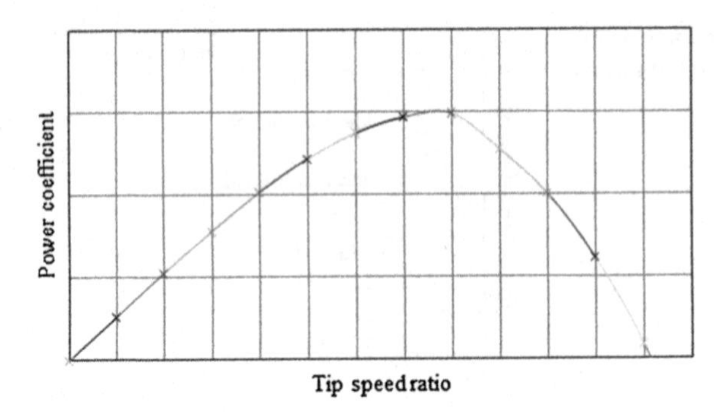

Fig. 1.10 Power coefficient in terms of TSR

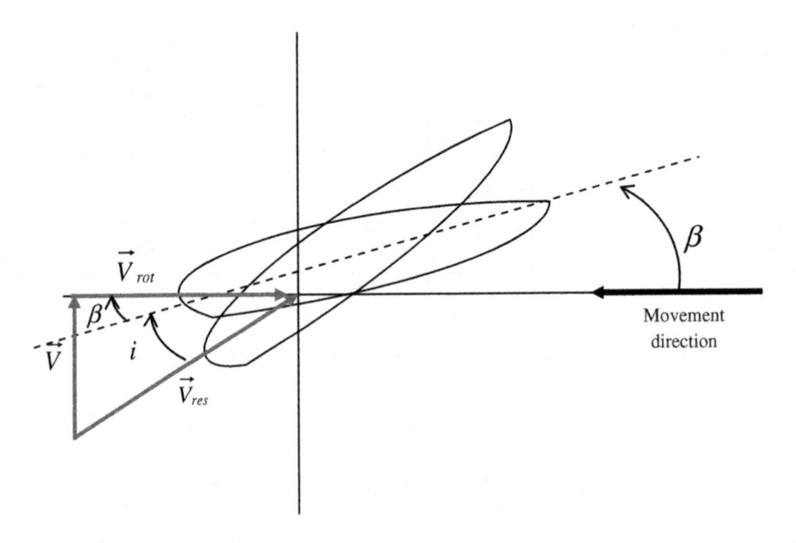

Fig. 1.11 Pitch control

In this case, the power coefficient can be expressed in terms of constant coefficients [3]:

$$C_p(\lambda) = c_1\left(\frac{c_2}{\lambda_i} - c_3 \cdot \beta - c_4\right) \cdot e^{\frac{c_5}{\lambda_i}} + c_6 \cdot \lambda \tag{1.7}$$

where c_1, c_2, c_3, c_4, c_5, and c_6 are constant coefficients (Table 1.3).

The expression (1.7) can then be written as follows:

$$C_p(\lambda) = 0.5176 \cdot \left(\frac{116}{\lambda_i} - 0.4 \cdot \beta - 5\right) \cdot e^{-\frac{21}{\lambda_i}} + 0.0068 \cdot \lambda \tag{1.8}$$

Table 1.3 Values of the coefficients c_1, c_2, c_3, c_4, c_5, and c_6

Coefficient	Value
c_1	0.5176
c_2	116
c_3	0.4
c_4	5
c_5	21
c_6	0.0068

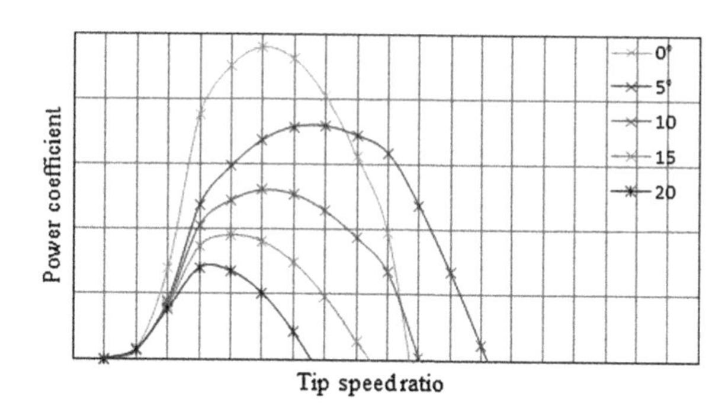

Fig. 1.12 Power coefficient for different pitch angle beta

where

$$\frac{1}{\lambda_i} = \frac{1}{\lambda + 0.008k \cdot \beta} - \frac{0.035}{\beta^3 + 1} \tag{1.9}$$

We can represent the power coefficient for different pitch angle beta as shown in Fig. 1.12.

- **We can have the simplified expression** (1.10) as follows (Fig. 1.13):

$$C_p(\lambda) = 0.5176 \cdot \left(\frac{116}{\lambda_i} - 0.4 \cdot \beta - 5\right) \cdot e^{-\frac{21}{\lambda_i}} \tag{1.10}$$

- **The power coefficient C_p is related to tip speed ratio λ and rotor blade pitch angle**:

$$C_p(\lambda, \beta) = (0.44 - 0.0167\beta) \sin\left(\frac{\pi(\lambda - 3)}{15 - 0.3\beta}\right) - 0.00184(\lambda - 3)\beta \tag{1.11}$$

We can represent the power coefficient for different pitch angle beta as shown in Fig. 1.14.

We can observe that for a fixed pitch angle β, a maximum power coefficient C_p is achieved when the TSR λ is at the optimal value.

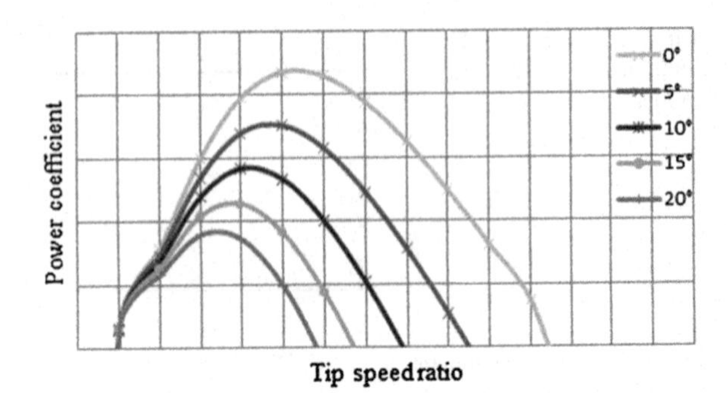

Fig. 1.13 Power coefficient for different pitch angle beta with the simplified expression

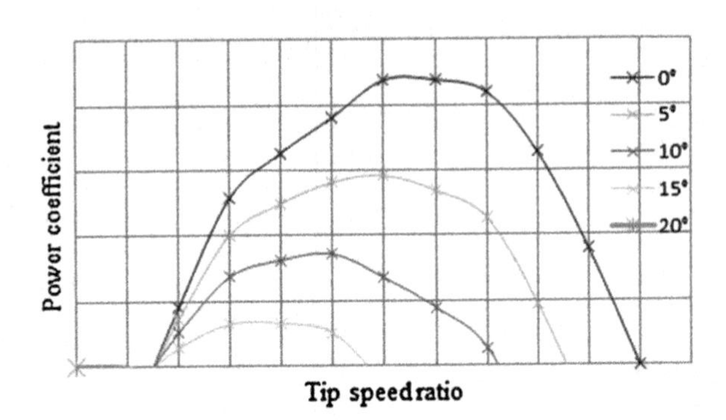

Fig. 1.14 Power coefficient as a function of TSR and pitch angle

- ***C_p* in an approximate relationship**

The following approximate relationship for C_p used:

$$C_p(\lambda, \beta) = \frac{1}{2} \cdot \left(\frac{R}{\lambda} - 0.022 \cdot \beta^2 - 5.6 \right) \cdot e^{-\frac{0.17.R}{\lambda}} \quad (1.12)$$

Power Coefficient in Terms of Wind Speeds

We express the power coefficient by an approximation function using polynomial interpolation [4]

$$\begin{aligned} C_p(V_{\text{wind}}) = {} & 1.1072 - 1.2698 * V_{\text{wind}} - 0.493 * V_{\text{wind}}^2 - 0.0008 * V_{\text{wind}}^3 + 0.0781 \\ & * V_{\text{wind}}^4 - 4.27 * \exp(-4) * V_{\text{wind}}^5 + 1.37 * \exp(-5) * V_{\text{wind}}^6 - 2.44 * \exp(-7) \\ & * V_{\text{wind}}^7 + 1.83 * \exp(-9) * V_{\text{wind}}^8 \end{aligned} \quad (1.13)$$

The power coefficient can be represented as follows (Fig. 1.15):

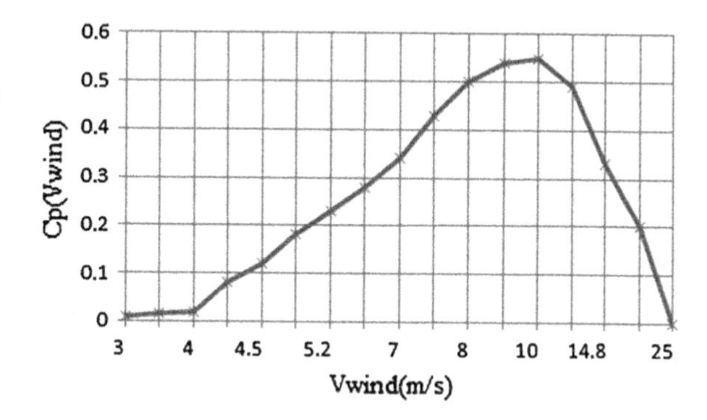

Fig. 1.15 Power coefficient obtained with an approximation function using polynomial interpolation

1.2.2.4 Wind Power Expression

The aerodynamic power of the wind through a wind disk of radius R is given by [5]:

$$P_{\text{wind}} = \frac{1}{2} \cdot \rho \cdot \pi \cdot R^2 \cdot V_{\text{wind}}^3 \tag{1.14}$$

where ρ represents the air density ($\rho = 1.235\ \text{kg/m}^3$) and $A_{\text{wind}} = \pi \cdot R^2$ is the swept area.

For example, for a small turbine ($R = 1$ m), we obtain the following curve (Fig. 1.16) for different wind speeds:

We can also express the wind power by an approximation function using polynomial interpolation [4]:

$$\begin{aligned} P_{\text{wind}} &= 4{,}240 - 4{,}727 * V_{\text{wind}} - 2{,}194 * V_{\text{wind}}^2 - 562 * V_{\text{wind}}^3 \\ &\quad + 88.5 * V_{\text{wind}}^4 - 8.91 * V_{\text{wind}}^5 + 0.585 * V_{\text{wind}}^6 \\ &\quad - 0.0249 * V_{\text{wind}}^7 + 6.64 * \exp(-4) * V_{\text{wind}}^8 \end{aligned} \tag{1.15}$$

We can represent it in Fig. 1.17:

We can deduce the aerodynamic torque expression by:

$$P_{\text{wind}} = \omega_{\text{tb}} \cdot T_{\text{wind}} \tag{1.16}$$

Thus,

$$Te_{\text{wind}} = \frac{1}{2} \cdot C_T(\lambda, \beta) \cdot \rho \cdot \pi \cdot R^3 \cdot V_{\text{wind}}^2 \tag{1.17}$$

where $C_T(\lambda, \beta)$ is the torque coefficient (TC) and it depends on the power coefficient $C_P(\lambda, \beta)$ and is defined as follows:

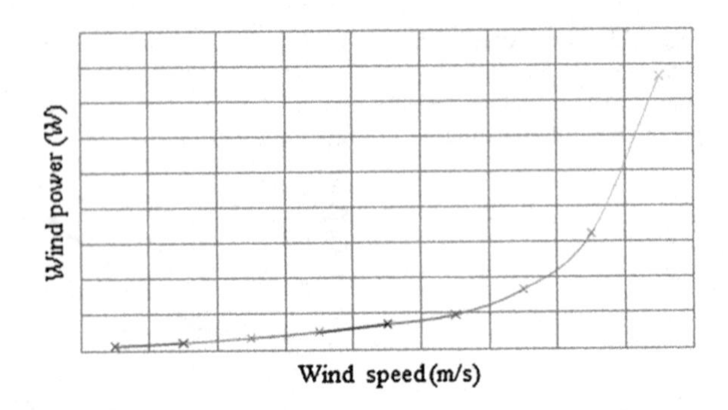

Fig. 1.16 Aerodynamic power versus wind speeds

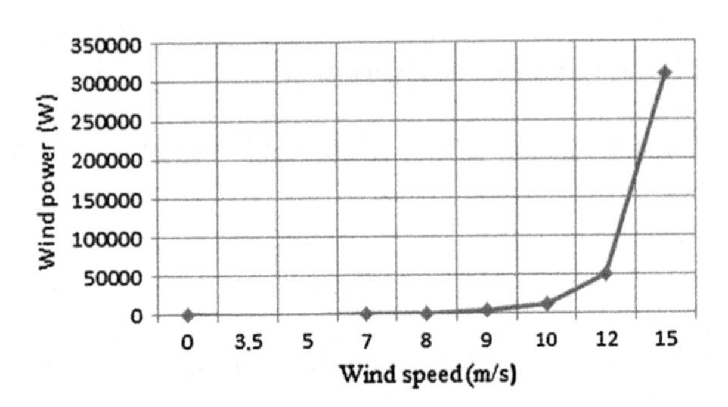

Fig. 1.17 Aerodynamic power obtained with polynomial interpolation

$$C_T(\lambda,\beta) = \frac{C_P(\lambda,\beta)}{\lambda} \tag{1.18}$$

The torque coefficient (TC) can be represented as Fig. 1.18.

The mechanical power, which is converted by a wind turbine, P_t, is dependent on the power coefficient $C_p(\lambda,\beta)$. It is given by:

$$P_t = \frac{1}{2}\cdot C_p(\lambda,\beta)\cdot\rho\cdot\pi\cdot R^2\cdot V_{\text{wind}}^3 \tag{1.19}$$

A wind turbine can only convert just a certain percentage of the captured wind power. This percentage is represented by $C_p(\lambda,\beta)$ which is a function of the wind speed, turbine speed, and the blade pitch angle β (Fig. 1.19).

The $C_P(\lambda,\beta)$ curve has a unique maximum $[C_P(\lambda,\beta)]_{\text{opt}}$ that corresponds to a maximum power. Where

$$\lambda_{\text{opt}} = \frac{\omega_{t-\text{opt}}}{v_{\text{wind}}} \tag{1.20}$$

The theoretical maximum power coefficient value is given by the Betz limit

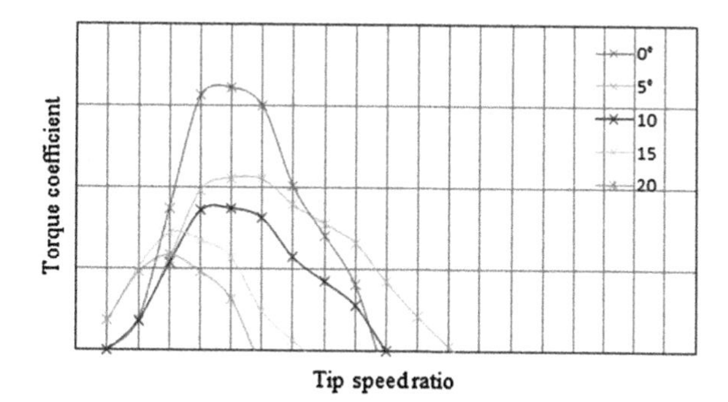

Fig. 1.18 Torque coefficient for different pitch angle beta

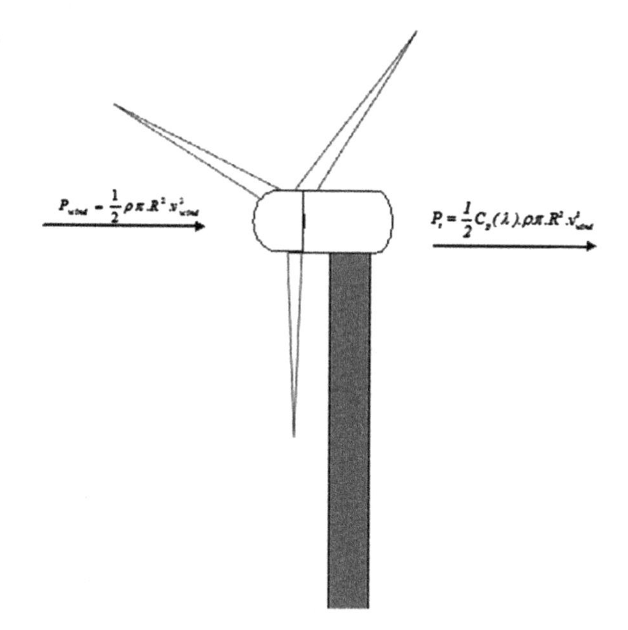

Fig. 1.19 Power conversion in a wind turbine

$$C_{P,\max} = \frac{16}{27} \approx 0.5926 \tag{1.21}$$

The theoretical maximum wind power will be:

$$P_{\max} = \frac{1}{2} \cdot \frac{16}{27} \cdot \rho \cdot \pi \cdot R^2 \cdot V_{\text{wind}}^3 \tag{1.22}$$

Thus (Fig. 1.20),

$$P_{\max} = 0.3644 * \pi \cdot R^2 \cdot V_{\text{wind}}^3 \tag{1.23}$$

We can also obtain the maximum power coefficient by the following expression [2]:

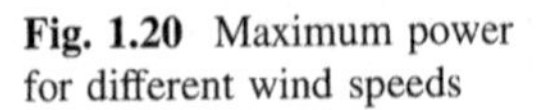

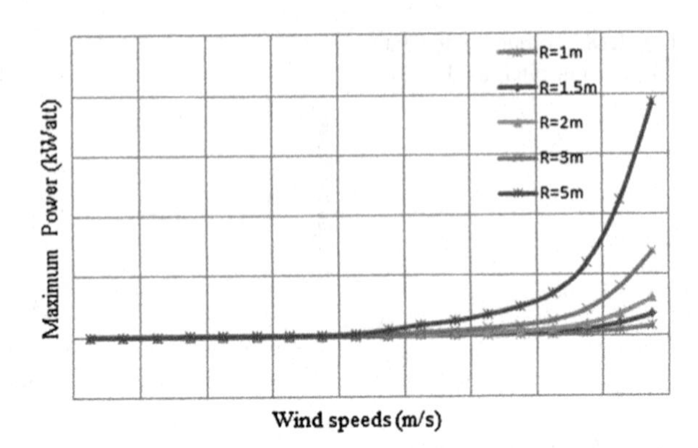

Fig. 1.20 Maximum power for different wind speeds

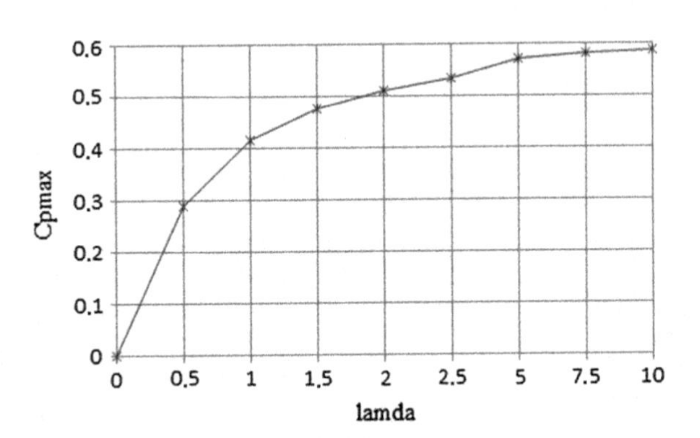

Fig. 1.21 Ideal theoretical maximum power coefficient

$$C_{p,\max} = \frac{24}{\lambda^2} \cdot \int_{a1}^{a2} \left[\frac{(1-a)\cdot(1-2a)\cdot(1-4a)}{(1-3a)} \right]^2 \cdot da \tag{1.24}$$

where a_1 corresponds to the axial induction factor for $\lambda = 0$ and a_2 corresponds to the axial induction factor for λ equal to the local speed ratio.

We can then obtain the curve Fig. 1.21.

1.2.2.5 Turbine Control

The turbine control includes two major parts: power control and pitch control. The combination of pitch control and electrical power control results in two distinct operating conditions (see Chap. 3).

- For wind power less than rated, the blades are fixed to capture the maximum wind power and turbine speed will be controlled by adjusting electrical power.
- For wind power above rated, the rotor speed will be controlled by the pitch for full power operation, with transient speed being allowed to rise above the reference.

1.2.2.6 Electrical Generator

There are different EGs for wind turbines. It depends principally on turbine concept.

Fixed-Speed Wind Turbine Systems

This operation case is used when the wind speed is controlled by pitch control. The fixed-speed wind system consists of a squirrel cage induction generator (SCIG) directly connected to the grid (Fig. 1.22).

The pitch control system of the blades kept constant the rotational speed of the machine, driven through a gearbox. This structure is the conventional concept applied during the decade 1980–1990. Besides the simplicity of this system, this solution has the advantages of the induction generator which is robust and standard. Moreover, there is a simple direct connection due to the slip variation occurring between the stator flux and rotor speed. However, the SCIG requires reactive energy to ensure the rotor magnetization. Generally, capacitors are connected in parallel with the generator to ensure the reactive power consumption (Fig. 1.23).

The advantages of fixed-speed wind turbines using SCIG are as follows [5]:

- Simple electrical system;
- High reliability;
- Moderate cost.

Their major disadvantages are as follows:

- The extracted power is not optimized: With this type of wind turbine, we have not the possibility of adjusting the power generated.
- Lack of reactive power management by induction generator: The direct connection of an induction generator to the grid requires the addition of capacitor banks to reduce reactive power required to the grid.

Variable-Speed Wind Turbine System

This system is very simple, but it can be noisy due to the orientation of the blades which limits the range of exploitable wind speeds. For this, a variable-speed operation will be necessary to maximize the power extracted from the wind. But in this case, a direct connection to the grid is not possible because of the stator voltage frequency variations. A power electronic (PE) interface between the generator and the grid is necessary. It consists of two converters (a rectifier and an inverter) connected via a DC voltage bus.

The main advantages of this configuration are turbine system based on SCIG [5]:

- Increase in energy efficiency;
- Reduction in torque ripple;
- Generation of an electrical power with a high quality.

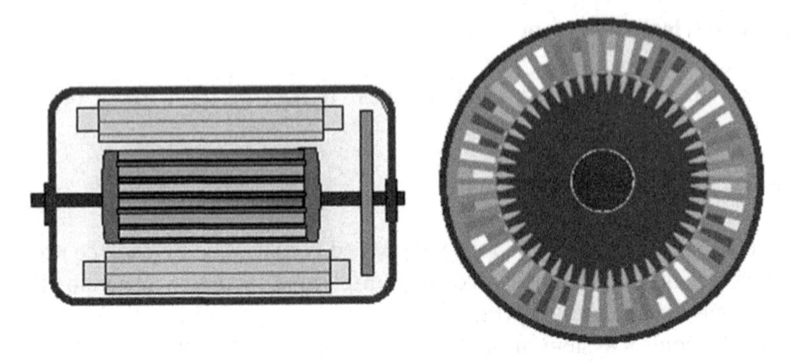

Fig. 1.22 Induction generator [5]

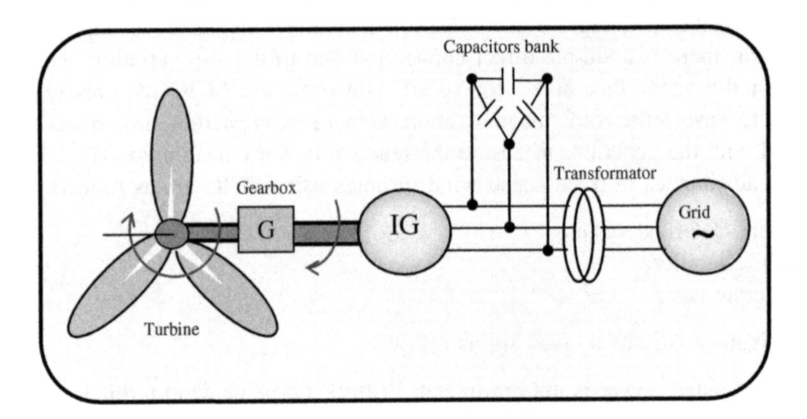

Fig. 1.23 Fixed-speed wind turbine system based on SCIG

The use of "complex" power converters is its main disadvantage.
In variable-speed wind turbine system, we can use different generators.

Squirrel Cage Rotor Induction Generator

Induction generator can be used in variable-speed wind turbine by introducing a frequency converter. This configuration allows a variable-speed operation, but it is generally expensive. Indeed, the frequency converter size must be 100 % of rated power stator of the electrical machine. For these reasons, this configuration is rarely used because it is not competitive with other types of machines, especially the doubly fed induction machine (DFIM) (Fig. 1.24).

Dual-Stator Induction Machine

A dual-stator induction machine (DSIM) is an induction machine which consists of a standard squirrel cage rotor and a stator with two separate wound windings for a dissimilar number of poles. Each stator winding is fed from an independent variable-frequency variable-voltage inverter (Fig. 1.25).

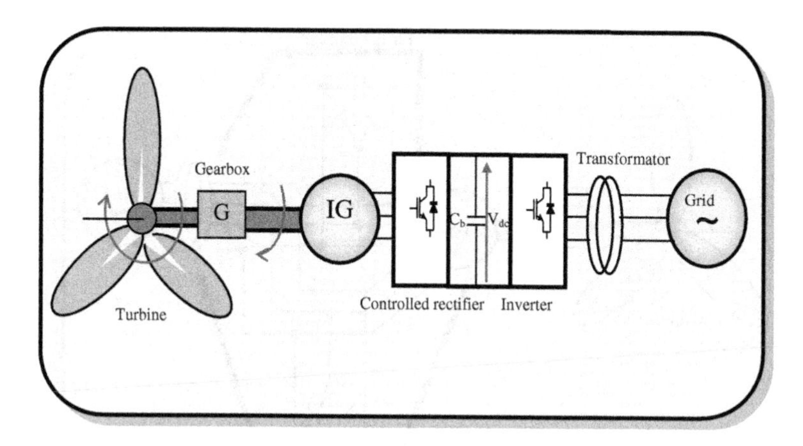

Fig. 1.24 Wind energy system based on squirrel induction machine with variable frequency

Doubly Fed Induction Generator

The doubly fed induction generator (DFIG) is a kind of induction machine in which both the stator windings and the rotor windings are connected to the source. With synchronous machines, it is currently one of the two competing solutions in variable-speed wind turbine. A DFIG used in wind turbines offers different advantages as operation at variable rotor speed, generation of electrical power at lower wind speeds, and control of the power factor. Wind conversion system consists of coupling the rotor of the DFIG to the grid through two three-phase PWM inverters. The first one operates in a rectifier mode; the second one operates in an inverter mode. In general, the size of the system is limited to 25 % of the rated stator power of the electrical machine, which is sufficient to assure a variation of 30 % of the speed range. That is its main advantage, while its main drawback is related to interactions with the grid, especially over currents caused by grid voltage [5].

Self-Cascaded Machine or Brushless Doubly Fed Machine

The brushless doubly fed machine (BDFM) requires a special rotor structure formed by several loops. It has its origins from the cascade technology induction machines, and it consists of two sets of three-phase windings with different number of pole pairs in the stator and a special rotor cage (Fig. 1.26).

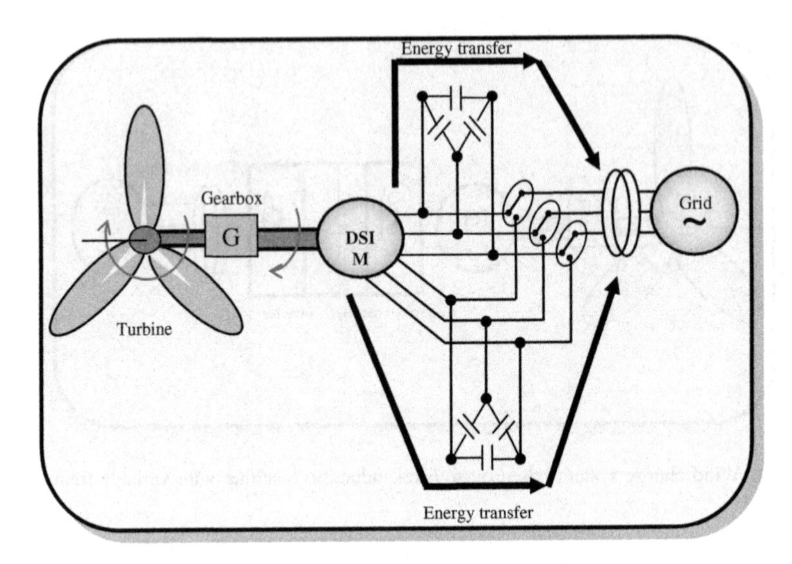

Fig. 1.25 Wind energy system based on dual-stator induction machine

The most advantages of this structure are as follows:

- The gearbox could be done with one single stage of reduction.
- The rated power of the converter is lower than the nominal power of the generator (this advantage similar to that of the DFIG).
- Robust machine has a great overload capacity, and installation in hostile atmospheres is easy.

Dual-Stator Winding Induction Machine

The dual-stator winding induction machine (DSWIM) has a normal squirrel-cage rotor design and two windings on the stator, one with high power and the other with low power. Therefore, it is different from the brushless doubly fed induction machine which has a special rotor design. This idea of the dual-stator winding induction machine design provides some advantages such as reducing the cost of the dual-stator winding induction machine. It has the same advantages as the synchronous machine connected to the grid via an electronic power interface [5] (Fig. 1.27).

Doubly Fed Induction Machine with Wound Rotor

In this case, the machine operates like a synchronous motor whose synchronous speed can be varied by adjusting the frequency of the AC currents fed into the rotor windings. So, mechanical power at the machine shaft is converted into electrical power supplied to the AC power grid via both the stator and rotor windings. Several structures have been developed using these machines, we can find [5].

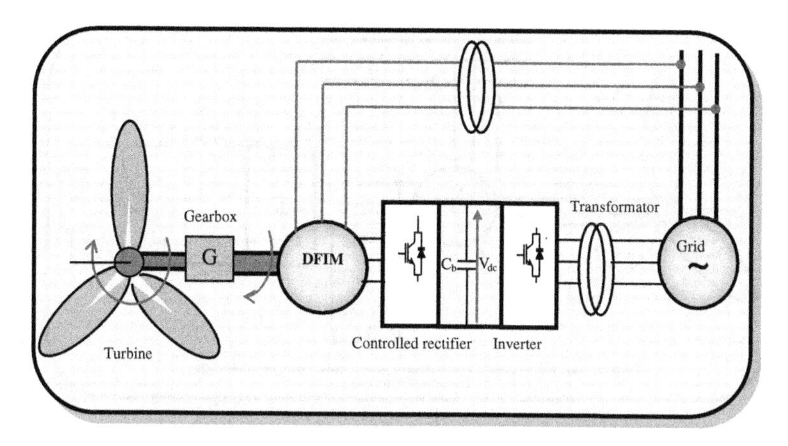

Fig. 1.26 Wind system based on a doubly fed induction motor with variable frequency

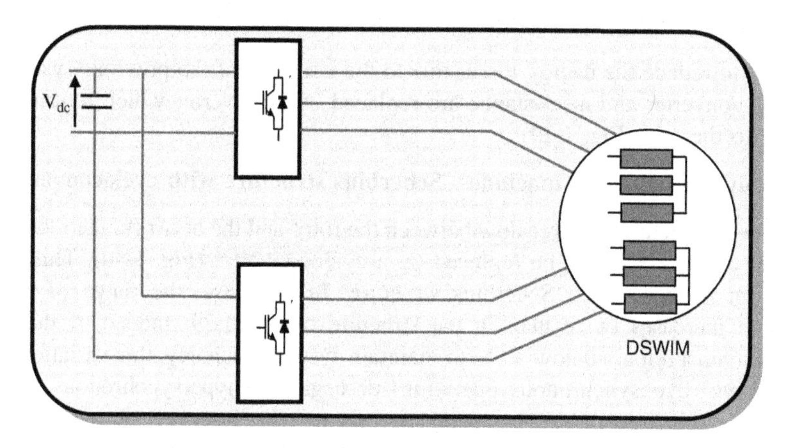

Fig. 1.27 Dual-stator winding induction machine (DSWIM)

- **The doubly fed induction machine with energy rotor dissipation**

In this configuration, the stator is connected directly to the grid and the rotor is connected to a rectifier. A resistive load is placed at the output of the rectifier through a DC/DC converter. The control of the DC/DC converter is used to vary the energy dissipated by the rotor winding and to obtain variable-speed operation in the stable part of the torque/speed characteristic of induction machine. Then, machine slip will vary with the rotational speed variations. For a constant slip, the power extracted from the rotor side is high and entirely dissipated in the resistance R, which fails to maintain system efficiency, so in this case, the dimensions of the converter and the resistance will be increased (Fig. 1.28).

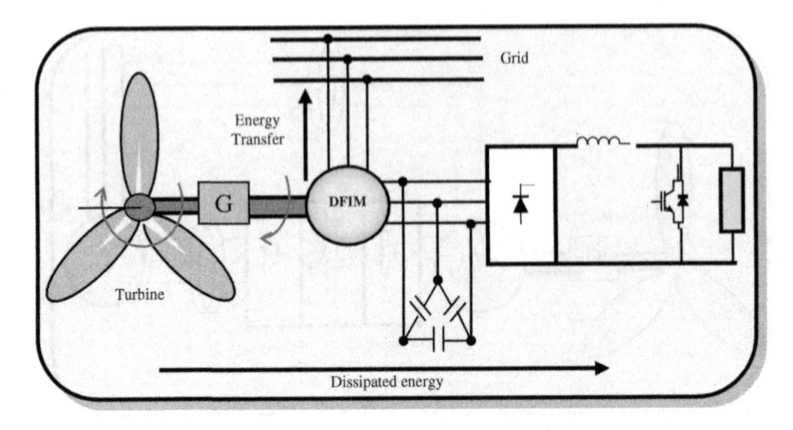

Fig. 1.28 Doubly fed induction machine with energy rotor dissipation

- **Doubly fed induction machine—Kramer structure**

In order to reduce the energy losses due to the structure of the previous system, a DC/DC converter and a resistance are replaced by an inverter which returns slip energy to the grid (Fig. 1.29).

- **Doubly fed induction machine—Scherbius structure with cycloconverter**

To allow bidirectional power flow between the rotor and the network, the rectifier–inverter combination can be replaced by a cycloconverter (Fig. 1.30). This configuration is then called Scherbius structure. In this case, the range of speed variation increases twice than in the structure of Fig. 1.29. Indeed, if the slip variation must remain below 30 % to maintain system efficiency, this variation can be positive (hyposynchronous operation) or negative (hypersynchronous operation). The principle of the cycloconverter is to take fractions of the network sinusoidal voltages to reproduce a lower frequency wave. Therefore, its use generates significant harmonic disturbances that affect the power factor of the device. Advances in PEs have led to the replacement of the cycloconverter by a two IGBT converters controlled by PWM.

- **Doubly fed asynchronous machine—Scherbius structure with PWM converters**

This configuration has the same characteristics as the Scherbius structure with cycloconverter. The bidirectional converter allows hyposynchronous and hypersynchronous operation and power factor control of the grid side (Fig. 1.31).

- **Doubly fed induction machine—Scherbius structure with "Matrix converters"**

The matrix converter (see Chap. 2) is an interesting alternative that overcomes the disadvantages of the previous structure. Indeed, it is characterized by [5]:

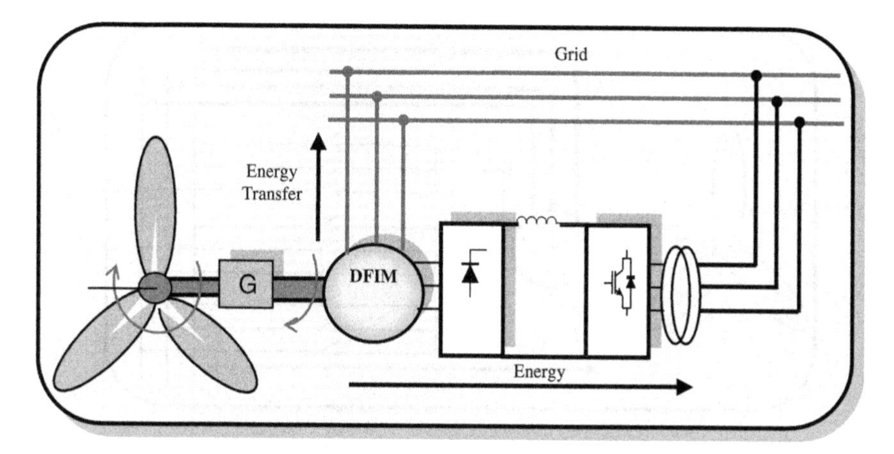

Fig. 1.29 DFIG with Kramer structure

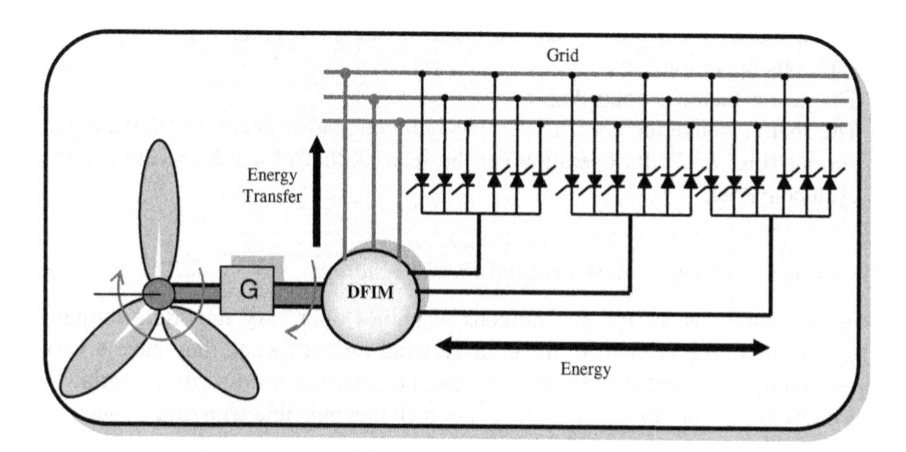

Fig. 1.30 Scherbius structure with cycloconverter

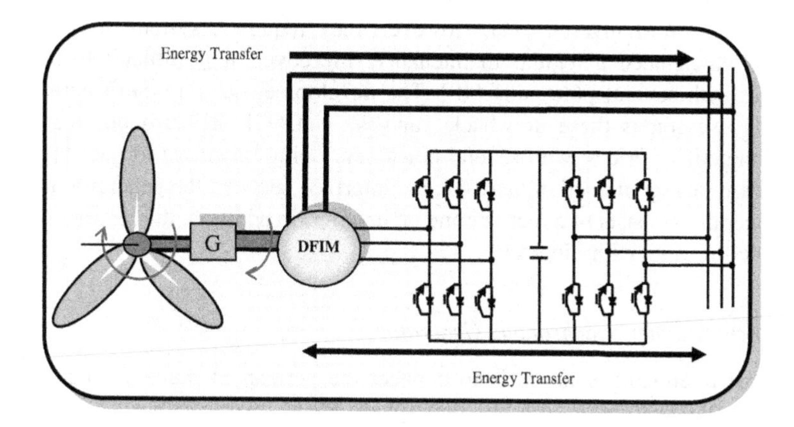

Fig. 1.31 Scherbius structure with PWM converters

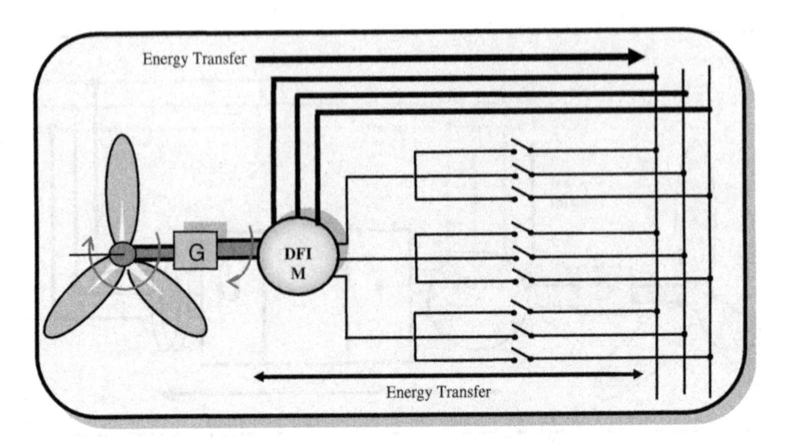

Fig. 1.32 Scherbius structure with matrix converters

- It is a bidirectional power converter, and it offers the possibility of hyposynchronous operation.
- The power factor is controllable.
- The matrix converter consists of nine bidirectional switches for voltage and current (Fig. 1.32). Its operation can be seen as that of a forced commutated cycloconverter.

Synchronous Generator with External Field Excitation

With the same size of IG, synchronous machines offer very important torques. They can therefore be used in direct-drive wind turbines when they have a very large number of poles [6–15]. In this case, operation is necessarily in variable-speed frequency and the stator frequency is then incompatible with grid frequency. We have to use converters [17]. Wind turbines based on a wound rotor synchronous generator (SG) are interesting, especially when field current is a parameter which can be useful for energy optimization and the armature current is controlled through the PWM inverter [17]. However, they require a system of rings and brushes which need a regular maintenance. Moreover, it is difficult to build a machine with several pole pairs [18]. The development of rare earth permanent magnets overcomes these drawbacks, and we can build synchronous machines with competitive prices, with several poles developing important torques [16]. As in the case of cage induction machine, the interface between the generator and the grid generally consists of a rectifier and an inverter. In addition, the rectifier can be controlled or not (with diodes).

Permanent Magnet Synchronous Generator

The most interesting solution is to connect the permanent magnet synchronous generator (PMSG) stator with two three-phase PWM inverters (Fig. 1.33). In this

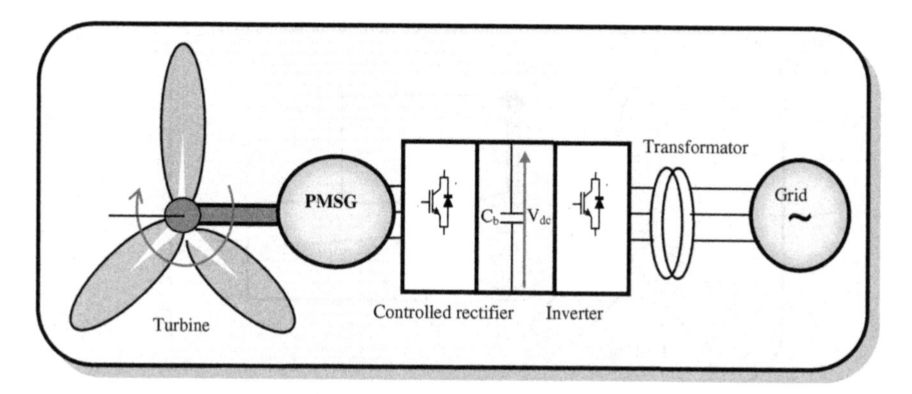

Fig. 1.33 Wind energy system based on permanent magnet synchronous machine with variable frequency

case, the interface with the grid can be completely controlled via the inverter, while the rectifier controls the power generated by the PMSG. Moreover, this type of configuration ensures a decoupling between the behavior of the wind generator and the grid behavior [18]. However, the two converters must be sized for the generator-rated power, which is the main drawback of this configuration. Also, structures, interposing a chopper between the rectifier and the inverter, are used. The chopper allows an indirect control of the transmitted power that permits an operation at the maximum power point tracking (MPPT).

Other Electrical Generators Used

Many other types of wind generators are also mentioned in literatures, such as:

- **Switched reluctance generators (SRG)** [9]

The principles of operation of SRG are based on reluctance torque. The machine has a stator of wound salient poles and develops a torque that tends to align the poles in a way that increases the reluctance in the magnetic circuit. We can use SRG in Turbine–generator direct coupling operation (Fig. 1.34a) or coupling to the turbine shaft through a gearbox (Fig. 1.34b). The instantaneous power available in a SRG is a function of the inductance L, the position of the rotor θ, the rotor speed ω_r, and the number of phases n_{phase} and can be expressed as follows:

$$P(\theta, k_j) = \left[\frac{1}{2}\sum_{k=1}^{n_{\text{phase}}} \frac{dL_k(\theta)}{d\theta} i_k^2\right] \cdot \omega_r \tag{1.25}$$

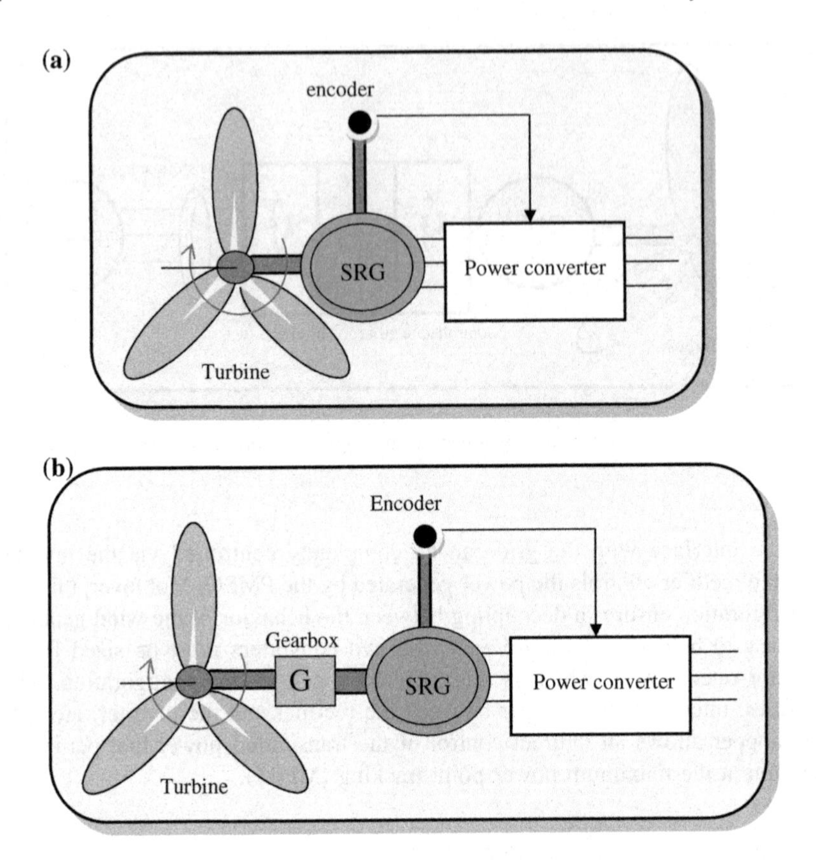

Fig. 1.34 Switched reluctance generator in a wind turbine. **a** Direct-drive wind turbine with SRG. **b** Indirect coupling with gearbox

The average resulting power can be derived from:

$$P = T_{\text{em}} \cdot \omega_r \tag{1.26}$$

where ω_r is the angular velocity and T_{em} is the average value of torque.

- **Linear induction generators (LIG)** [5]

A linear motor has its stator and rotor "unrolled" so that, instead of producing a torque, it produces a linear force along its length. The force is produced by linear magnetic field acting on conductors in the field. Any conductor, which is placed in this field, will have eddy currents induced in it, thus creating an opposing magnetic field due to Lenz's law. The two opposing fields will repel each other, thus creating motion as the magnetic field sweeps through the metal [16].

- **Claw-pole generators** [5]:

Claw-pole generator (CPG) is a special type of wound rotor SG (Fig. 1.35).

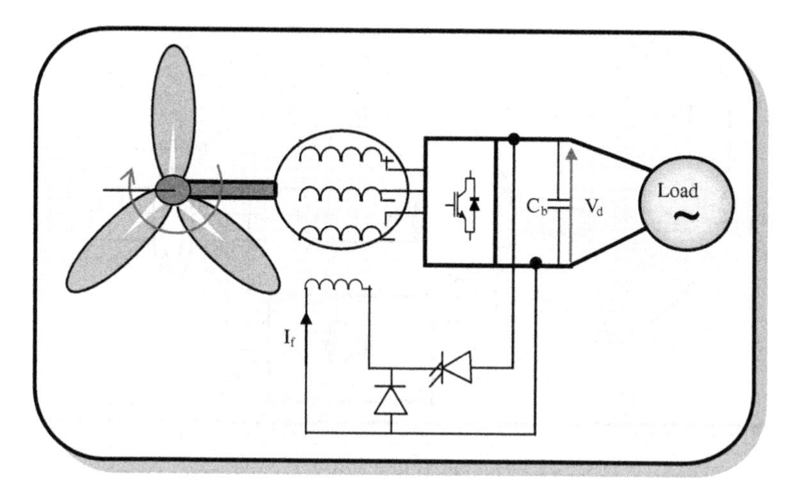

Fig. 1.35 Claw-pole generator

- **Transverse-flux machines**

Compared to classical motor concepts, a completely different mechanical structure is necessary. It can be with PMTFM or without permanent magnet excitation. PMTFM has an achievable torque density two times larger than that of conventional or tooth wound winding permanent magnet synchronous motors [17].

- **Brushless DFIGs (BDFIGs)**:

The BDFIG requires double-stator windings, with different number of poles in both stator layers [5]. The first stator is connected to the grid, and thus, the generator output frequency must be equal to the grid frequency. The second stator winding is connected through a power converter, which is rated at only a fraction of the wind turbine rating (Fig. 1.36).

This configuration allows controlling the stator active and reactive power. The advantage of this system is that this concept does not require slip rings. Its drawback is its complexity in machine operation principle and its assembly.

1.2.3 Power Electronics Interface

The type of electric generator used and the load/grid conditions dictate the requirements of the PE interface (see Chap. 3).

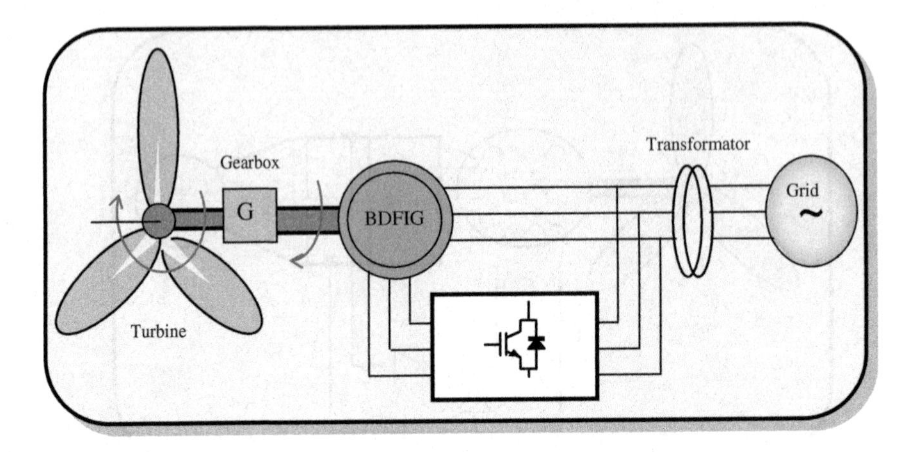

Fig. 1.36 Wind energy system based on brushless doubly fed induction generator

1.2.4 Load

The load can be an autonomous load or a grid utility.

1.2.4.1 Isolated or Autonomous Load

Even all attention is on powerful machines, it is observed an increasing demand for smaller units to be installed near homes or buildings to use electrical energy directly. Induction generator is widely used for the production of electricity from wind energy, especially in remote and isolated areas. With all its advantages (simplicity, robustness, low maintenance, small size per kW generated), it is the generator the most used in the production of low power in isolated or autonomous operation.

The "small wind" covers the power range from 20 to 100 kW with three categories [11]:

- micro wind turbines from 20 to 500 W,
- mini wind turbines from 500 to 1 kW,
- small wind turbines with 1 to 100 kW.

In the case of micro and mini wind turbines, the power output is less than 1 kW, so it is necessary to use a lower cost system which can ensure the highest performances. In the case of small wind turbines, we obtain a good energy and a good operation of the global system. So, it is necessary to add a PE interface to feed loads with fixed voltages and frequencies. Different methods had been proposed in different researches.

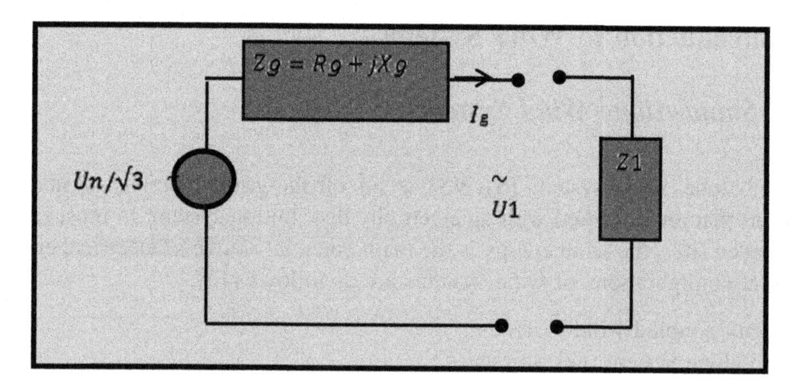

Fig. 1.37 Equivalent scheme

1.2.4.2 Utility Grid

We can identify two types of wind turbines connected to the grid:

- fixed-speed wind turbines which consist of an induction machine with squirrel cage
- variable speed wind turbines with of a double-fed induction machine (DFIG) or a permanent magnet synchronous machine (PMSM).

Basic Grid Properties

It can be approximated by a first-order equivalent scheme (Fig. 1.37).

The apparent short-circuit power can be calculated as follows:

$$S = \sqrt{3} \cdot U_n \cdot I_g \tag{1.27}$$

Grid Connection Requirements

One major requirement at present time is the connection and optimized integration of large wind farms into electrical grids. With increased wind power capacity, transmission system operators (TSOs) have become concerned about the impact of high levels of wind power generation on power systems. Thus, TSOs have issued grid codes and grid requirements for wind turbines connection and operation. The main issues of grid codes can be summarized as follows [15]:

- Active power control,
- Reactive power control,
- Voltage and frequency control,
- Power quality, for example flickers and harmonics,
- Fault ride-through capability.

1.3 Introduction to Wind Systems

1.3.1 Stand-Alone Wind Systems

A stand-alone wind system (SAWS) is an off-the-grid electricity system for locations that are not fitted with an electricity distribution system. In most remote and winded sites, the wind energy is the main potential source of electrical energy. Different configurations of wind systems are as follows [15]:

- Directly coupled wind system
- Stand-alone system with storage
- Hybrid system.

1.3.2 Directly Coupled Wind System

We present the various solutions of wind energy conversion applications used in these low-power systems for both isolated and autonomous systems.

1.3.2.1 Synchronous Machines

Wound Rotor Synchronous Machine

This type of machine uses generally a current field excitation which requires the presence of a power supply. Therefore, isolated sites are suitable for these generators only if there is a battery or a voltage source [16].

Permanent Magnet Synchronous Machines

The PMSM is a very interesting solution in autonomous wind energy applications due to its advantages (a good efficiency and a good mass torque), and there is no need for a power source for the excitation circuit. Its main drawback is its high cost than induction machines. However, different structures of PMSMs supplying loads through autonomous PE devices exist. We will give a brief presentation in Fig. 1.38.

Diode Rectifier Structure

This configuration is the simplest one and finds its applications in the case of very-low-power systems [9]. It is based on the direct association of a battery with the bridge rectifier diodes (Fig. 1.39). In this case, there are no controlled components, nor are few sensors, and then, the system cost is minimal. The operation is

Fig. 1.38 PMSG connected to a wind turbine

"natural" but requires a very precise choice of all parameters (voltage and machine parameters) by a system design [16].

Structure with Diode Rectifier Bridge and Chopper

The WCES should enable operation at maximum power to maximize energy efficiency, regardless of the wind variations. This is the principle of MPPT. However, with only the association of a diode rectifier bridge with a permanent magnet SG, it is not always possible to achieve MPPT. In order to obtain this, a buck chopper is placed after the diode bridge, which acts as a storage battery (Fig. 1.40) [19, 21].

Bridge Structure with a Differential Control

In the same view for optimizing energy efficiency, a bridge structure with a differential control allows operation with a duty cycle close to 0.5 by controlling the two switches T_1 and T_2 (Fig. 1.41). This configuration is advantageous in terms of operating reliability but requires two more components, resulting in a higher cost and therefore higher losses [21].

Structure with PWM Rectifier

The reference configuration is obviously this one (Fig. 1.42) where the generator is connected to a three-phase PWM rectifier. It is possible to obtain a dynamical control which makes it easy to move the operating point over the entire speed rotation range. But, it requires a more complex system; a complete three arms with six switches, and a control unit that normally requires a mechanical position sensor.

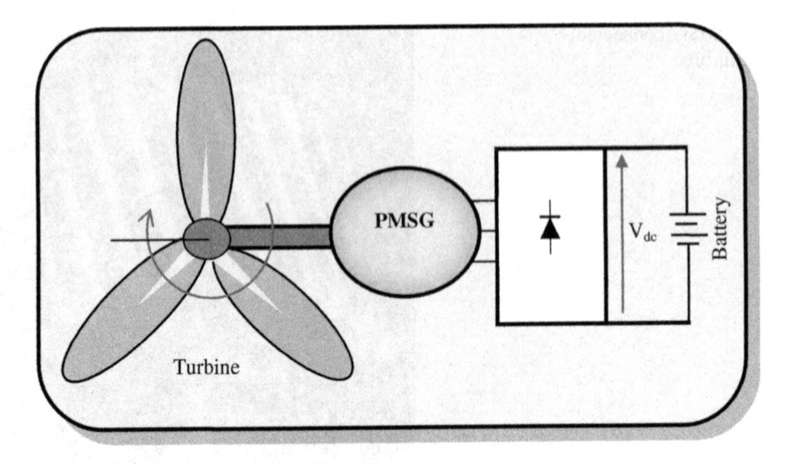

Fig. 1.39 Synchronous machine with diode rectifier

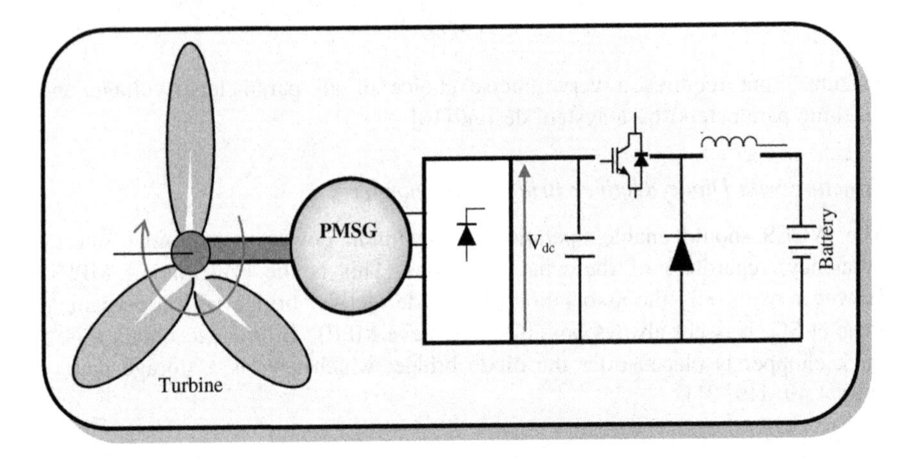

Fig. 1.40 Synchronous machine connected to a diode rectifier bridge and buck chopper

Variable Reluctance Machines

Some works have also involved in the use of variable reluctance synchronous machine for application in wind energy conversion [21–23]. It has the structure of stator windings which are similar to those of an induction machine. Its rotor may be provided with a squirrel cage, which ensures the direct starting on the grid and improves the stability of operation in synchronism. As the induction motor, reluctance synchronous motor absorbs reactive power. The power factor is relatively low, which limits its application area to a few tens of kilowatts [17]. However, the cage dampers are not obligatory, and the manufacturing cost can be very attractive to use in autonomous wind turbine. The generator operation of this

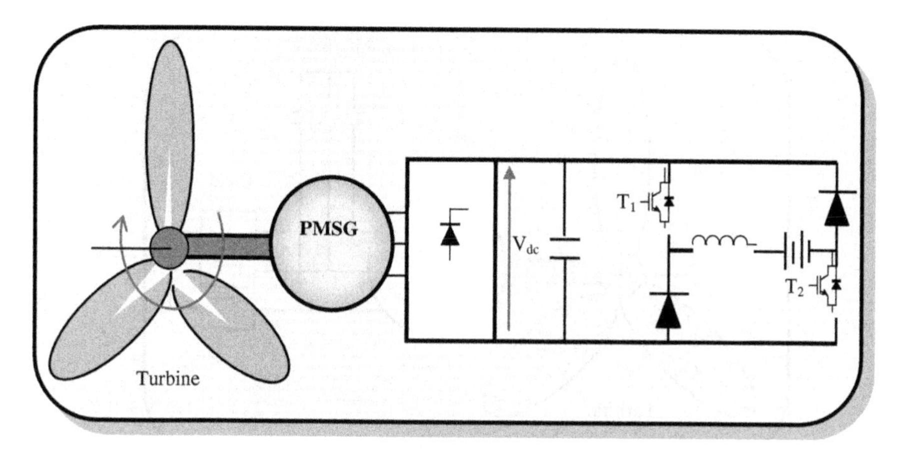

Fig. 1.41 Synchronous machine with converter diode bridge and chopper

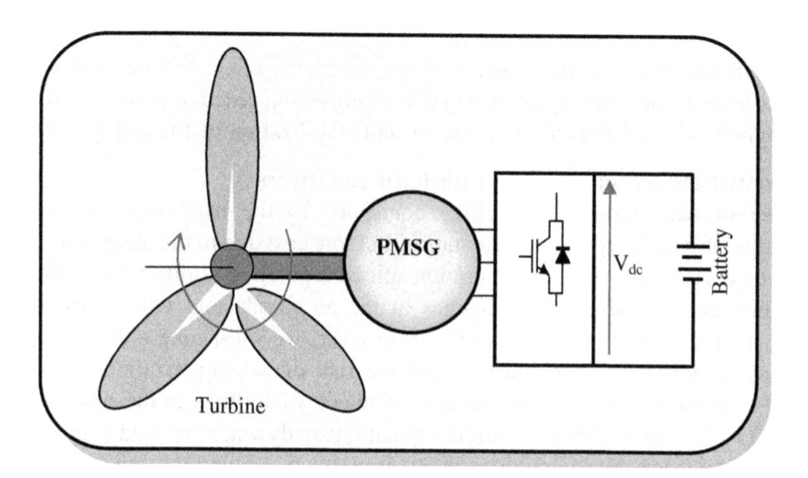

Fig. 1.42 Synchronous machine connected to a PWM rectifier

structure is very close to that of a squirrel induction machine. Therefore, it can be used in stand-alone wind system feeding a bank of capacitors in parallel with the load (Fig. 1.43) or on a PWM rectifier [23].

Induction Machines with Squirrel Cage

Squirrel induction machine is the one of the most widely used machines for application in stand-alone WCES due to its robustness and price. The maintenance cost is well below that of an alternator of the same power [21]. Finally, it is

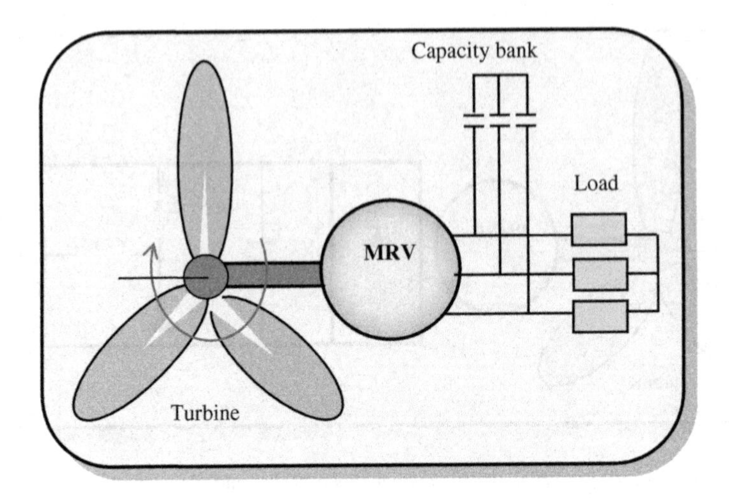

Fig. 1.43 Variable reluctance synchronous machine connected to bank capacitors

tolerant in extreme operations (over speed, overload, etc.). In what follows, we give a summary of a few systems capable of providing reactive power needed for the magnetization of induction generator and stabilization of the stator voltage.

Self-excited induction machine with fixed capacitors

The use of cage induction machine connected to the bank capacity has the advantage of simplicity in implementation and low cost due to the absence of static converters. However, this configuration allows operating limits which must be taken into account, especially changes in the amplitude and voltage frequency during load and speed variations [19]. Moreover, we must take precautions and avoid overloading the machine to avoid the risk of demagnetization. There are several configurations in the literature to connect capabilities to the machine terminals. One of those is to use a single capacity, usually when we feed single-phase loads. In this case, one enough capacity is used, the generator can provide the necessary power to the load, which reduces the total system cost. However, in case of default phase where the capacitor is connected, or in the capacitor itself, the machine is demagnetized in the absence of another source of reactive power. Also, this structure can induce unbalanced stator currents. Another solution is to use a bank capacity of three capacitors connected in star or delta at the terminals of the generator. This ensures more dependability for system wind generation, and the load is single or three phase. In case of failure in one phase or in a capacitor, the system can operate, but there will be a voltage drop due to the decrease in the magnetization of the generator. This configuration can also lead to unbalanced stator currents in the case of supplying a single load phase or three-phase unbalanced load. In what follows, we describe the different configurations.

Self-excited induction machine by a single fixed capacitor
Several works have been carried out in the case where the induction generator is self-excited by a single excitation capacity. The winding of the machine is connected in star or delta structures.

In case where the machine is connected in star, the capacity excitation can be connected between one phase and neutral as it can be connected between two phases.

The first case is treated in [20–22], and each author provides a way of connecting the load:

- The load is in parallel with the capacitor excitation [21, 22].
- The load is between neutral and phase where capacity is not connected [21].
- The load is between the other two phases where the capacity is not connected [20].

The second case is treated in [20, 21]. The first author proposes the connection of the load in parallel with the capacity of excitation. But, the second author proposes the connection.

- The first is to connect a capacitor and two other charges, each between two terminals of the three phases of the machine.
- In the second, one connects a capacitor in parallel with a phase of the machine and a load between the other two phases.

In case where the machine is connected in delta, the structures shown in Fig. 1.44 have been proposed in [23]. The capacity of excitation is connected between two terminals. The load can be placed in parallel with the capacity (Fig.1.44a) or between one of two terminals, where the capacity is connected, and the remaining terminal of the triangle (Fig. 1.44b).

Self-excited induction machine by a fixed battery bank
In this configuration, the stator windings of the induction machine are connected to a capacity bank in parallel with the load. Many studies have been devoted to this structure as shown in Fig. 1.45 [23–35]. In other works, the connection of capacitors is in delta [36, 37].

Self-excited asynchronous machine by a fixed capacitor with a compensation system
The use of a fixed capacity bank does not control the flow of reactive power and therefore do not maintain the amplitude and frequency of the voltage supplied which is constant during load variations or wind speed. To overcome this disadvantage, while keeping a capacity bank for self-excitation, several alternatives were considered.

- **Short-shunt connection**

The most commonly structure used is to connect, in addition to parallel capacitors, other capacitors in series with the load [23, 38] or with the stator of the machine [39] (short-shunt connection). This approach reduces the voltage drop in load [18], but it restricts the possibility of a continuous control voltage over a wide range of

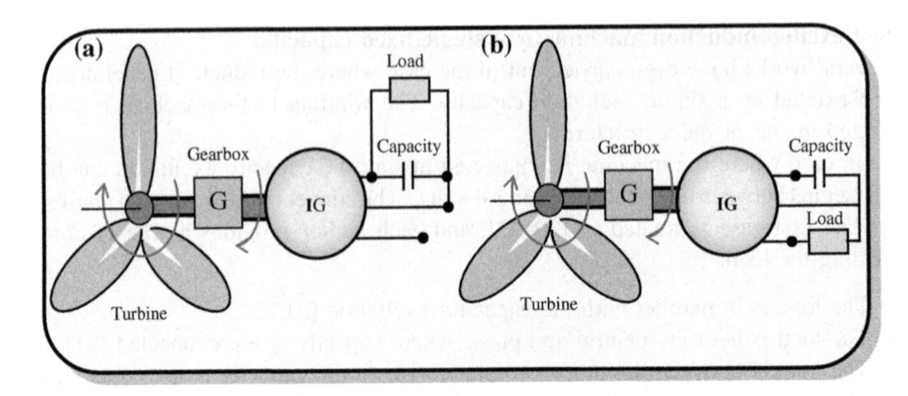

Fig. 1.44 Self-excited asynchronous machine with a single capacitor. **a** Capacity of excitation connected between two terminals and **b** Capacity between one of two terminals

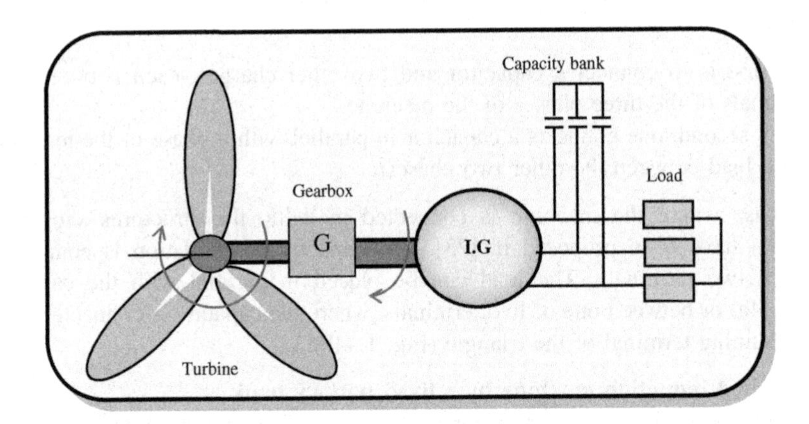

Fig. 1.45 Self-excited induction machine with a capacitor bank

loads and/or speeds [40]. Figure 1.46 shows the self-excited induction machine with a capacitor bank and compensation where capacity is in series with the load.

- **Controlled capacity bank**

Another solution, based on the concept of continuous control of the excitation battery, was proposed in [35]. This is a device consisting of fixed capacities in parallel with switches of gate thyristor off (GTO), connected in anti-parallel (see Fig. 1.47). The apparent value of the capacity can then be adjusted periodically by controlling the time during which the capacitor is connected to the circuit.

This device operates as a variable capacitor, so as to create a source of variable reactive energy which allows decreasing the voltage variations when disturbing the wind speed or load. However, in case of total discharge of the storage device, it cannot obtain energy production.

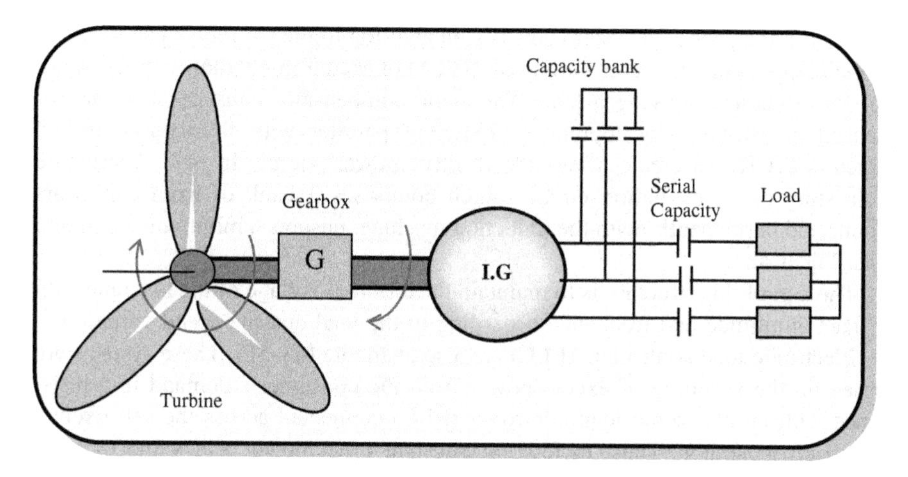

Fig. 1.46 Self-excited machine with short-shunt connection

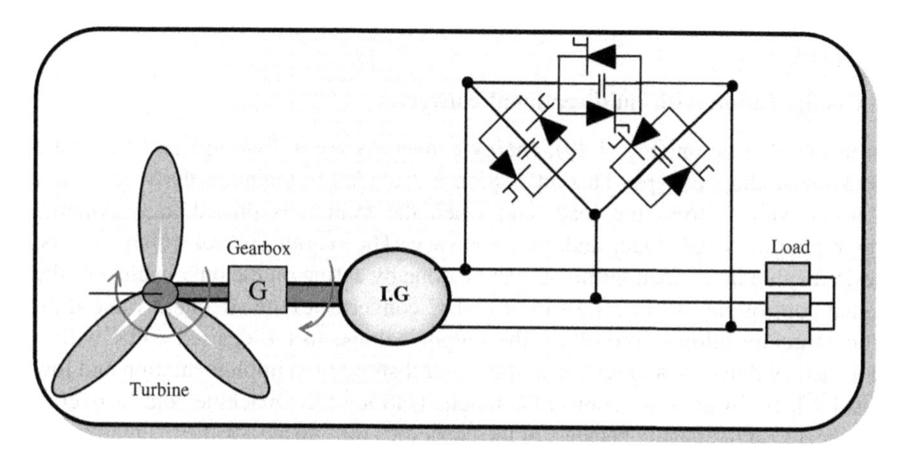

Fig. 1.47 Self-excited induction machine with a capacitor bank and graduators

- **Generator reactive power**

More effective solutions, but also more complex, have been proposed as static reactive power generators (static VAR generator). They use systems based on PEs [35]. Several structures of static compensators of reactive power called static VAR compensator (SVC) have been proposed to maintain the voltage constant:

1. Thyristor-controlled reactor (TCR) [41].
2. Thyristor-switched capacitor (TSC) [41].
3. The static compensator (STATCOM) [42–44].
4. The voltage source inverter (VSI) [45–49].

In [41], the proposed system consists, in addition to the parallel capacities, of a static compensator for reactive power (SVC) connected in parallel with the stator of the machine supplying a load. The static compensator consists of a bank of capacitors switched by thyristors (TSC), in parallel with thyristor-controlled reactors (TCR), to create a variable reactive power source. In [47], a structure consisting of an excitation circuit which comprises a bank of fixed capacitors connected permanently with the induction machine, ensures a minimum excitation (see Fig. 1.48).

The aim of this structure is to maintain the terminal voltage of the machine with a fixed amplitude and frequency according to the load and the speed variations.

Electronic load controllers (ELC) are also available [48–52]. These systems are based on the shedding of excess power from the consumer's demand in a resistance with an electronic load controller (ELC) connected across the self-excited induction generator. Thus, in [52], at constant input power and value of fixed capacity, the induced voltage varies with the applied load. IGBT switches are used to control the connection and the disconnection of the load dump. All systems based on PEs previously cited provide satisfactory results in terms of voltage regulation, but their disadvantages are the complexity of implementation and high costs [40].

- **Configuration with unidirectional converter**

In this configuration (Fig. 1.49), energy converters are a PD3-type rectifier and a buck/boost chopper type. This adaptation is intended to maintain the voltage at a constant value across the load, and when the system is placed in a dynamic environment, speed, load, and capacity vary. The output control voltage of the desired value in the load terminals can be done by acting on the duty cycle α of the signal controlling the switch of the DC–DC converter to the adequate value [52]. This structure allows controlling the output voltage to the desired value with a satisfactory dynamic and with a simple control strategy in implementation and low cost [48]. But it does not control the magnetization of the machine, and its overall cost keeps on increasing because of the capacities introduced for the magnetization and the use of two converters and a filtering system.

Another solution is given in [54], where the load is connected after the LC filter (without the DC interface). It allows reducing the overall cost compared to the previous configuration, but it does not control the voltage across the load and the magnetization of the machine. In [31], the rectifier is composed of two different components which are diodes and thyristors and the load is connected directly to the output of the rectifier (without DC–DC interface and the LC filter). This solution also reduces the overall cost compared with the previous structure and the control of the voltage across the load, but it does not allow control of the machine magnetization.

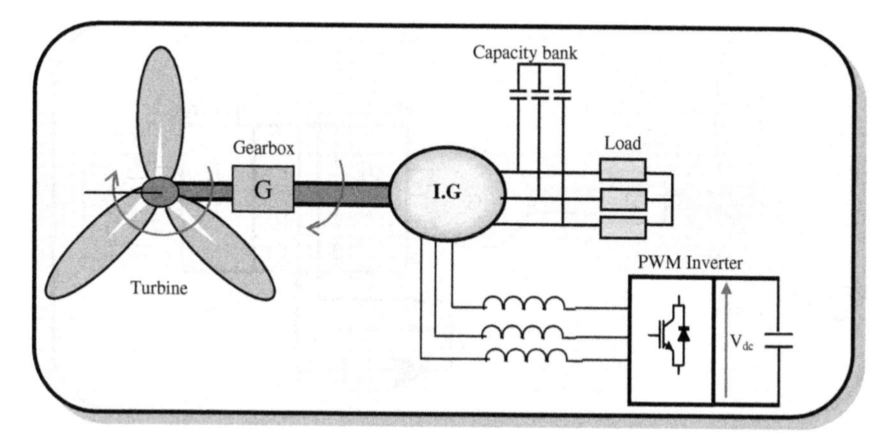

Fig. 1.48 Structure with PWM converter in parallel with the load [53]

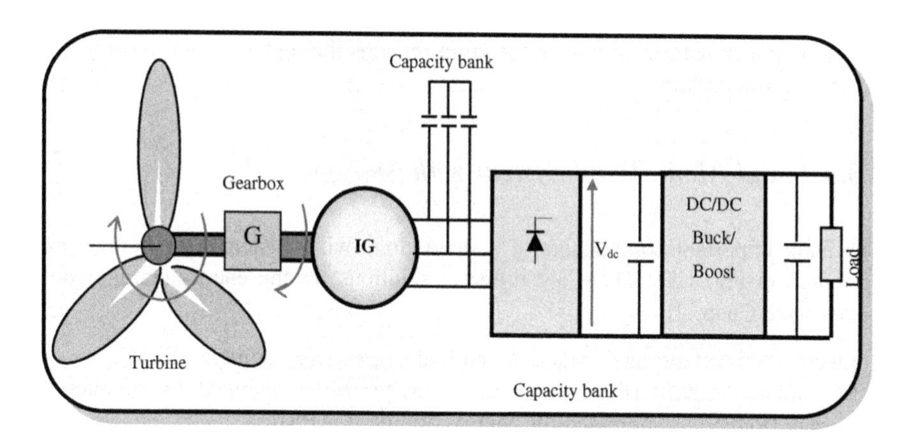

Fig. 1.49 Structure with diode bridge and chopper

- **Configuration with bidirectional converters**

The structure shown in Fig. 1.50 consists of an induction generator connected to a rectifier and a PWM inverter. This configuration allows the control of the magnetization and the voltage at the output of the induction generator when the rotor speed and the electrical load vary [55–64].

The insertion of static converters between the generator and its load allows obtaining two new degrees of freedom. Others, which are used in the case of a suitable control, lead to a better exploitation of the wind, which can result in the following advantages [22]:

- Operation at low speeds.
- A reduction in acoustic noise.
- An optimization of energy transfer.
- A good management of transients toward the load.

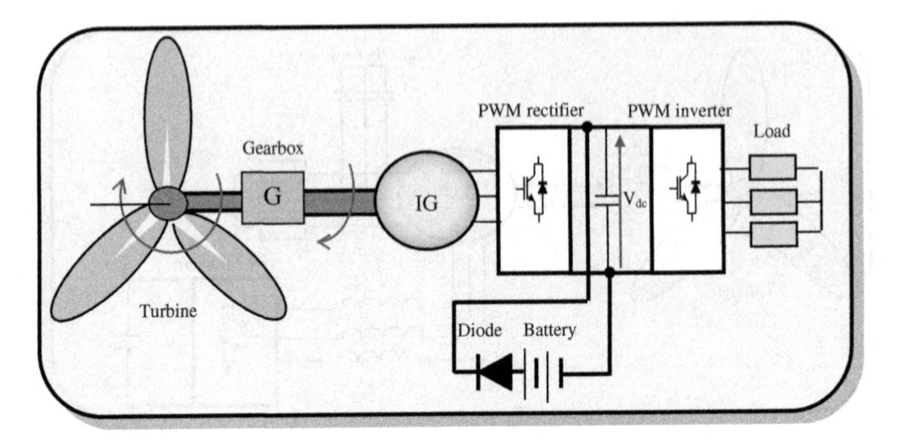

Fig. 1.50 Structure based on two PWM converters

The major drawback of this architecture remains the high cost and complexity of the control strategy.

1.3.3 Stand-Alone Wind System with Storage

The main important components of a stand-alone wind system with storage are presented in Fig. 1.51. There are different solutions for the energy storage subsystem (see Chap. 4):

- Electrochemical storage [batteries and hydrogen energy storage (HES)],
- Mechanical storage [flywheel energy storage (FES), pumped hydro energy storage (PHES), compressed air energy storage (CAES)],
- Electromagnetic storage energy [supercapacitor energy storage (SES), superconducting magnetic energy storage (SMES)]
- Others (see Chap. 4).

1.3.4 Hybrid System

Hybrid power systems (HPSs) combine two or more sources of renewable energy as one or more conventional energy sources [65–78]. The renewable energy sources such as photovoltaic and wind do not deliver a constant power, but due to their complementarities, their combination provides a continuous electrical output. HPSs are generally independent from large interconnected networks and are often used in remote areas. The purpose of a hybrid power system is to produce as much energy from renewable energy sources to ensure the load demand. In addition to sources of energy, an hybrid system may also incorporate a distribution system to DC distribution system AC, a storage system, converters, fillers, and an option

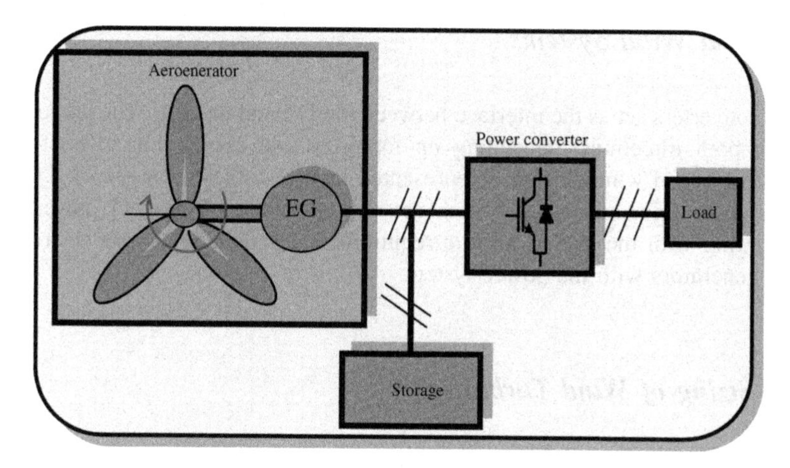

Fig. 1.51 Main components of a stand-alone wind system with storage

to load management or supervision system. All these components can be connected in different architectures. The renewable energy sources can be connected to the bus or DC, depending on the size of the system. The power delivered by HPS can vary from a few watts for domestic applications up to a few megawatts for systems used in the electrification of small villages. Thus, for the hybrid systems used in applications with very low power (under 5 kW), they are generally used to feed DC loads. Larger systems, with a power greater than 100 kW, are connected to the AC bus and are designed to be connected to large interconnected networks [69]. Hybrid systems are characterized by several different sources, several different loads, several storage elements, and several forms of energy. Wind generators in a hybrid system can be connected in three configurations, DC bus architecture, AC bus architecture, and DC–AC bus architecture.

1.3.4.1 Advantages and Disadvantages of Stand-Alone Wind System

Advantages of Hybrid System

- Not dependent on one source of energy.
- Simple to use.
- Efficiency, low cycle cost of living component of the hybrid system.

Disadvantages of a Hybrid System

- More complex than single-source systems and the need for storage
- high capital cost compared to diesel generators

1.3.5 Grid Wind Systems

Power converters act as the interface between the EG and the grid. The power may flow in both directions, depending on topology and applications. The typical scheme of a grid wind system is represented in Fig. 1.52. It consists of a power generation, a grid, and loads, which are the customers. Delivered power synchronization with the grid is a basic requirement for interconnecting distributed power generators with the power system.

1.3.6 Sizing of Wind Turbine

Sizing an isolated wind energy system is different from sizing a system for gird integration. But generally, it depends mainly on the site location that dictates

- the average wind speed,
- the turbine orientation,
- the average energy consumption of the application.

1.3.6.1 Determination of Load Profile

The average energy consumption is determined through the application and the number of hours of their operation. For example, we display the load profile in Fig. 1.53.

1.3.6.2 Analysis of Wind Velocity

The average wind speed is obtained from the record of the meteorological station in the site location. For example, we present in Fig. 1.54 the average wind speed.

1.3.6.3 Calculation of Wind Energy

Mechanical power is given by Eq. 1.14. Then, wind energy produced P_{wind} by wind generator during a time Δt is given by:

$$E_{\text{wind}} = P_{\text{wind}} \cdot \Delta t \tag{1.28}$$

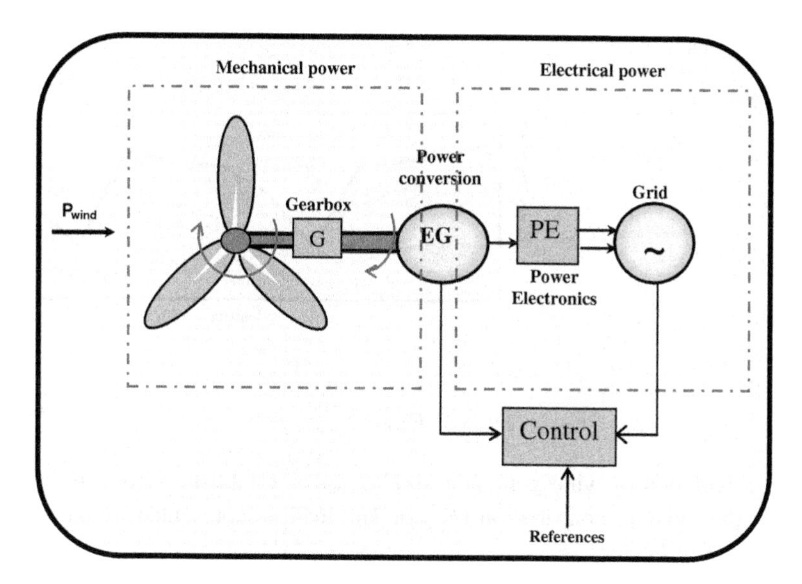

Fig. 1.52 Basic power conversion wind turbine system connected to the grid

Fig. 1.53 Example of a load profile during a summer day

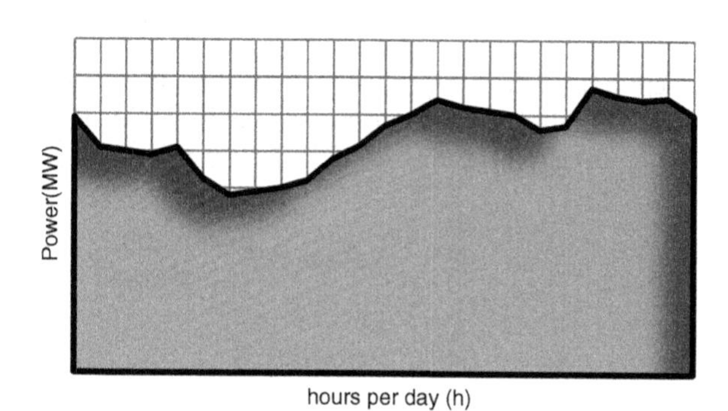

1.3.6.4 Size of Wind

The minimum surface of the generator needed to ensure full (100 %) coverage load (E_{Load}) is expressed by [75]:

$$S_{\text{wind},m} = \frac{E_{\text{Load},m}}{E_{\text{wind},m}} \tag{1.29}$$

where $E_{\text{wind},m}$ (kWh/m^2) is the monthly energy produced by the system per unit area and $E_{\text{Load},m}$ is the monthly energy required by the load (where m = 1, 2, …, 12 represents the month of the year).

The total energy produced by the wind generator which supplies the load can be expressed by:

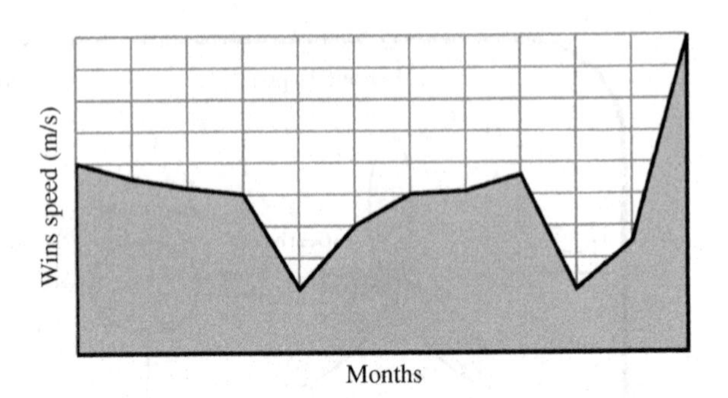

Fig. 1.54 Example of the average wind variations during a year

$$E_{\mathrm{Load}} = E_{\mathrm{wind}} \cdot S_{\mathrm{wind}} \tag{1.30}$$

The calculation of wind generator size (S_{wind}) is established from the annual mean of the monthly contribution ($\overline{E_{\mathrm{wind}}}$). The load is represented by the annual average energy $\overline{E_{\mathrm{Load}}}$.

$$S_{\mathrm{wind}} = f \cdot \frac{\overline{E_L}}{\overline{\overline{E_w}}} \tag{1.31}$$

where f is the fraction of load supplied by the wind energy.

The number of wind generator is calculated using the surface of the system unit $S_{\mathrm{wind},u}$, taking the entire value:

$$N_{\mathrm{wind}} = \mathrm{ENT}\left[\frac{S_{\mathrm{wind}}}{S_{\mathrm{wind},u}}\right] + 1 \tag{1.32}$$

1.3.6.5 Size of Storage

Battery Storage

Always, before calculations, we start by identifying the electrical usage. In the first step, we have to know the amount of energy we will be consuming per day $E_{\mathrm{Load,max}}$ (Wh/day). In the second step, we have to identify days of autonomy N_j (backup days). We multiply $E_{\mathrm{Load,max}}$ by this factor N_j.

$$E_{\mathrm{auto}} = E_{l,\max} \cdot N_j \tag{1.33}$$

Then, we have to identify depth of discharge (DOD) and convert it to a decimal value. Divide Eq. 1.30 by this value (DOD).

$$C = \frac{E_{\mathrm{auto}}}{\mathrm{DOD}} = \frac{E_{l,\max} \cdot N_j}{\mathrm{DOD}} \quad (1.34)$$

We have to associate each ambient temperature of a battery bank with coefficient of temperature factor FT. We select the multiplier corresponding to the lowest average temperature that batteries will be exposed to. That multiplier depends on the battery type (Table 1.4 gives an example of such data).

We multiply Eq. 1.34 by this factor (FT), and then, we obtain the minimum capacity of battery bank (Wh).

$$C_{\mathrm{batt,min}}(\mathrm{Wh}) = C \cdot \mathrm{FT} \quad (1.35)$$

Finally, we divide the minimum capacity of battery bank by battery voltage V_{batt} and we obtain the minimum capacity (Ah) of the battery bank.

$$C_{\mathrm{batt,min}}(A \cdot h) = \frac{C_{\mathrm{batt,min}}(W \cdot h)}{V_{\mathrm{batt}}} = \frac{E_{L,\max_j} \cdot N_j \cdot FT}{V_{\mathrm{batt}} \cdot DOD.}. \quad (1.36)$$

where V_{batt} is the battery voltage and DOD is the depth of discharge.

The number of batteries to be used is determined from the capacity of a battery unit $C_{\mathrm{batt},u}$ and is given by:

$$N_{\mathrm{batt}} = \mathrm{ENT}\left[\frac{C_{\mathrm{batt,min}}}{C_{\mathrm{batt},u}}\right] \quad (37)$$

Other Storage Systems

See Chap. 4.

1.4 Maintenance of Wind Systems

Gearbox and EG are the two most costly maintenance items for a wind system. In wind farms, generally an integrated online condition monitoring system can serve as an effective tool for managing day-to-day maintenance routines for a wind turbine and consolidating risky, costly maintenance activities [79]. Generally, there are two types of maintenance.

1.4.1 Large Maintenance

We have to make a full tightening of all bolts of the machine, lubrication, electrical testing, change filters, etc.

Table 1.4 Factor FT calculation [75]

Temperature in °C	Temperature in °F	Factor (FT)
+26	80+	1.00
+21	70	1.04
+15	60	1.11
+10	50	1.19
+4	40	1.30
−1	30	1.40
−6	20	1.59

1.4.2 Low Maintenance

We have to do some partial tightening, some electrical tests, lubrication, etc. In addition, annual maintenance of high voltage equipment and yearly inspection of pales must be made.

1.5 Total Costs for Wind Turbine Installation

Total costs for wind turbine installation depend on different parameters [80]:

- The number of turbines ordered,
- Financing cost,
- Construction contracts,
- The location of the project,
- Wind resource assessment and site analysis expenses,
- Construction expenses,
- Interconnection studies,
- Utility system upgrades,
- Transformers, protection, and metering equipment,
- Insurance, operations, warranty, maintenance and repair, legal and consultation fees,
- Taxes and incentives.

1.6 Onshore and Offshore Wind Power Technologies

1.6.1 Onshore Wind Power Technologies

Generally, we use the bladed, stall- or pitch-regulated, horizontal axis machine operating at near-fixed rotational speed. We can also use gearless "direct-drive" turbines with variable-speed generators. Control methods are pitch control and stall control.

1.6.2 Offshore Wind Power Technologies

Offshore turbines are taller and have longer blades, which results in a larger swept area and therefore higher electricity output.

1.7 Conclusion

This chapter is intended as an introduction to WCESs. It defines the wind process, introduces the main meteorological elements and the wind velocity, and presents an overview of stand-alone systems and grid-connected systems. We have also included short details on presizing, maintenance, and total costs of wind systems.

References

1. Schubel PJ, Crossley RJ (2012) Wind turbine blade design. Energies 5:3425–3449. doi:10.3390/en5093425
2. Akon AF (2012) Measurement of axial induction factor for a model wind turbine. Thesis submitted to the College of Graduate Studies and Research in partial fulfillment of the requirements for the degree of Master of Science in the Department of Mechanical Engineering University of Saskatchewan
3. Manwell JF, McGowan JG, Rogers AL (2010) Wind energy explained: theory, design and application. Wiley, London
4. Rekioua D, Abdelli R, Rekioua T, Tounzi A (2013) Modeling and control of an induction generator wind turbine connected to the grid, 2013 15th European Conference on Power Electronics and Applications, EPE 2013:1–6
5. Idjdarene K (2010) Contribution à l'étude et la commande de génératrices asynchrones à cage dédiées à des centrales électriques éoliennes autonomes. PhD, University of Bejaia-USTL de Lille
6. Hoffmann R (2002) A comparison of control concepts for wind turbines in terms of energy capture. Dissertation, Darmstadt
7. Libert F (2004) Design, optimization and comparison of permanent magnet motors for a low-speed direct-driven mixer. Partial fulfillment of the requirements for the degree of Technical Licentiate. Royal Institute of Technology Department of Electrical Engineering Electrical Machines and Power Electronics, Stockholm
8. Mittal R, Sandhu KS, Jain DK (2010) An overview of some important issues related to wind energy conversion system (WECS). Int J Environ Sci Dev 1(4):351–363
9. Bansal RC, Bhatti TS, Kothari DP (2002) On some of the design aspects of wind energy conversion systems. Int J Energy Convers Manage 43(16):2175–2187
10. Huynh QM, Nollet F, Essounbouli N, Hamzaoui A (2012) Fuzzy control of variable speed wind turbine using permanent magnet synchronous machine for stand-alone system. Sustain Energy Build Smart Innov Syst Technol 12:31–44
11. Tahour A, Aissaoui A, Essounbouli N, Nollet F (2012) Variable speed drive of wind turbine based on synchronous generator. Sustain Energy Build Smart Innov Syst Technol 12:3–16
12. Stiebler M (2008) Wind energy systems for electric power generation. Springer, Berlin

13. Wang L, Singh C, Kusiak A (2010) Wind power systems applications of computational intelligence. Springer, Berlin
14. Muyeen SM (2012) Wind energy conversion systems. Springer, Berlin
15. Erlich I, Bachmann U (2005) Grid code requirements concerning connection and operation of wind turbines in Germany. IEEE Power Eng Soc General Meet 2:1253–1257
16. Rekioua D, Rekioua T, Idjdarene K, Tounzi A (2005) An approach for the modeling of an autonomous induction generator taking into account the saturation effect. Int J Emerg Electr Power Syst 4(1):1–25
17. Rahim YHA (1993) Excitation of three-phase induction generator by a single capacitor. IEE Proc 140(1):1–7
18. Wang L, Cheng C (2000) Excitation capacitance required for an isolated three-phase induction generator supplying a single-phase load. IEEE Power Eng Soc Winter Meet 23–27:299–303
19. Wang L, Cheng C (2000) Selection of magnetization curves for accurately simulating a three-phase self-excited induction generator feeding a single-phase load. IEEE Power Eng Soc Winter Meet 23-27:286–290
20. Al-Bahrani AH, Malik NH (1990) Steady state analysis and performance characteristics of a three-phase induction generator self excited with a single capacitor. IEEE Trans Energy Convers 5(4):725–732
21. Wang L, Deng R (2006) A novel analysis of an autonomous three-phase delta-connected induction generator with one capacitor. IEEE Power Eng Soc General Meet 1–6
22. Elhafyani ML, Zouggar S, Benkaddour M, Zidani Y (2006) Permanent and dynamic behaviours of self-excited induction generator in balanced mode. Moroccan Stat Phys Soc 7(1):49–53
23. Ibtiouen R, Benhaddadi M, Nesba A, Mekhtoub S, Touhami O (2002) Dynamic performances of a self-excited induction generator feeding different static loads. In: Proceedings of 15th international conference on electrical machine ICEM 2002, Brugge, Belgium, 25–28 Aug 2002, pp 1–6
24. Kishore A, Kumar GS (2006) A generalized state-space modeling of three phase self-excited induction generator for dynamic characteristics and analysis. In: 1st IEEE conference on industrial electronics and applications, 24–26 May 2006, pp 1–6
25. Kishore A, Kumar GS (2006) Dynamic modelling and analysis of three phase self-excited induction generator using generalized state-space approach. In: Proceedings of international symposium on power electronics, electrical drives, automation and motion (SPEEDAM'06), IEEE, pp 52–59
26. Malik NH, Al-Bahrani AH (1990) Influence of the terminal capacitor on the performance characteristics of a self excited induction generator. IEE Proc 137(2):168–173
27. Nejmi A, Zidani Y, Naciri M (2002) Investigation on the self-excited induction generator provided with a hydraulic regulator. FIER, Tome II, Tétouane, Maroc, 8–10 Mai 2002, pp 494–499
28. Nesba A, Ibtiouen R, Touhami O (2006) Dynamic performances of self-excited induction generator feeding different static loads. Serbian J Electr Eng 3(1):63–76
29. Poitiers F, Machmoum M, Zaim ME, Branchet T (2002) Transient performance of a self-excited induction generator under unbalanced conditions. In: Proceedings of 15th international conference on electrical machine ICEM 2002, Brugge, Belgium, 25–28 Aug 2002, pp 1–6
30. Wang L, Kuo SC (2002) Steady state performance of a self-excited induction generator under unbalanced load. IEEE Power Eng Soc Winter Meet 1(27-31):408–412
31. Poitiers F, Machmoum M, Zaim ME, Branchet T (2002) Transient performance of a self-excited induction generator under unbalanced conditions. In: Proceedings of 15th international conference on electrical machine ICEM 2002, Brugge, Belgium, 25–28 Aug 2002, pp 1–6

32. Wang YJ, Huang SY (200) Analysis of a self-excited induction generator supplying unbalanced loads. In: Proceedings of international conference on power system technology (POWERCON'04), IEEE, pp 1457–1462
33. Shridhar L, Singh B, Jha CS, Singh BP, Murthy SS (1995) Selection of capacitors for the self-regulated short shunt SEIG excited induction generator. IEEE Trans Energy Convers 10(1):10–17
34. Bim E, Szajner J, Burian Y (1989) Voltage compensation of an induction generator with long-shunt connection. IEEE Trans Energy Convers 4(3):526–530
35. Al-Saffar M, Nho E, Lipo TA (1998) Controlled shunt capacitor self-excited induction generator. In: IEEE industry applications conference, thirty-third IAS annual meeting, vol 2. pp 1486–1490, 12–15 Oct 1998
36. Ahmed T, Noro O, Hiraki E, Nakaoka M (2004) Terminal voltage regulation characteristics by static VAR compensator for three-phase self-excited induction generator. IEEE Trans Ind Appl 40(4):978–988
37. Perumal BV, Chatterjee JK (2006) Analysis of a self-excited induction generator with STATCOM/battery energy storage system. In: IEEE power India power conference, pp 1–6
38. Singh B, Murthy SS, Gupta S (2006) STATCOM-based voltage regulator for self-excited induction generator feeding nonlinear loads. IEEE Trans Ind Electron 53(5):1437–1452
39. Singh B, Madhusudan M, Verma V, Tandon AK (2006) Rating reduction of static compensator for voltage control of three-phase self-excited induction generator. In: 2006 IEEE international symposium on industrial electronics, vol 2. pp. 1194–1199
40. Bellini A, Franceschini G, Lorenzani E, Tassoni C (2008) Quantitative design of active control for self excited induction generators in grid isolated operation. In: Proceedings of IEEE power electronics specialists conference (PESC'2008), IEEE, 15–19 Jun 2008, pp 3610–3614
41. Ahmed T, Nishida K, Nakaoka M (2007) Advanced control for PWM converter and variable-speed induction generator. IET Electr Power Appl 1(2):239–247
42. Lopes LAC, Almeida RG (2000) Operation aspects of an isolated wind driven induction generator regulated by a shunt voltage source inverter. In: Proceedings of IEEE industrial applications conference, conference record of the 2000 IEEE, vol 4. 8–12 Oct 2000, pp 2277–2282
43. Lopes LAC, Almeida RG (2006) Wind-driven self-excited induction generator with voltage and frequency regulated by a reduced-rating voltage source inverter. IEEE Trans Energy Convers 21(2):297–304
44. Timorabadi HS (2006) Voltage source inverter for voltage and frequency control of a stand-alone self-excited induction generator. In: Proceedings of IEEE CCECE/CCGEI, Ottawa, Canada, May 2006, pp 2241–2244
45. Singh B, Murthy SS, Gupta S (2004) Analysis and implementation of an electronic load controller for a self-excited induction generator. IEE Proc-Gener Distrib 151(1):51–60
46. Singh B, Murthy SS, Gupta S (2005) Transient analysis of self-excited induction generator with electronic load controller (ELC) supplying static and dynamic loads. IEEE Trans Ind Appl 41(5):1194–1204
47. Ramirez JM, Torres ME (2007) An electronic load controller for the self-excited induction generator. IEEE Trans Energy Convers 22(2):546–548
48. Elhafyani ML, Zouggar S, Aziz A, Benkaddour M (2007) Conception et modélisation d'un système éolien contrôlé par un régulateur de tension. In: Colloque international Sur les Énergies Renouvelables (CER'2007), Oujda, Maroc, 4–5 Mai 2007, pp 1–5
49. Kuo SC, Wang L (2002) Analysis of isolated self-excited induction generator feeding a rectifier load. IEE Proc Gener Trans Distrib 149(1):90–97
50. Idjdarene K, Rekioua D, Rekioua T, Tounzi A (2010) Performance of an isolated induction generator under unbalanced loads. IEEE Trans Energy Convers 25(2):303–311
51. Idjdarene K, Rekioua D, Rekioua T, Tounzi A (2008) Vector control of autonomous induction generator taking saturation effect into account. Energy Convers Manage Elsevier Sci 49(10):2609–2617

52. Levi E, Liao YW (1999) Rotor flux oriented induction machine as a DC power generator. In: Proceedings of 8th European conference on power electronics and applications EPE'99, EPFL Lausanne, Switzerland, pp 1–8
53. Rekioua D, Idjdarene K, Rekioua T, Tounzi A (2005) Vector controlled strategy of an autonomous induction generator: modeling and simulation. In: Proceedings of international conference on modeling and simulation (ICMS'05), Marrakech, Maroc, 22–24 Nov 2005, pp 1–5
54. Rekioua D, Idjdarene K, Rekioua T, Tounzi A (2007) Vector control strategy application to stand alone induction generator. In: Proceedings of international conference on electrical engineering design and technologies, Hammamet, Tunisie, 4–6 Nov 2007, pp 1–6
55. Sastry J, Ojo O, Wu Z (2006) High performance control of a boost AC-DC PWM rectifier-induction generator system. IEEE Trans Ind Appl 42(5):1146–1154
56. Idjdarene K, Rekioua D, Rekioua T, Tounzi A (2007) Control strategies for an autonomous induction generator taking the saturation effect into account. 2007 European Conference on Power Electronics and Applications, EPE:1–6
57. Mcgowan JG, Manwell JF (1999) Hybrid/PV/diesel system experiences. Rev Renew Energy 16:928–933
58. Belhamel M, Moussa S, Kaabeche A (2002) Production of electricity by means of a hybrid system (wind-photovoltaic-diesel), (traducted from French: Production d'Electricité au Moyen d'un Système Hybride (Eolien-Photovoltaïque-Diesel)). Renew Energy J 49–54
59. El Khadimi A, Bachir L, Zeroual A (2004) Sizing optimization and techno-economic energy system hybrid photovoltaic–wind with storage system. Renew Energy J 7:73–83
60. Kaldellisa JK, Kavadiasa KA, Koronakis PS (2007) Comparing wind and photovoltaic stand-alone power systems used for the electrification of remote consumers. Renew Sustain Energy Rev 11:57–77
61. Aly AEMMM, El-Aal A (2005) Modeling and simulation of a photovoltaic fuel cell hybrid system. A dissertation in candidacy for the Degree of Doctor in Engineering (Dr.-Ing.), Faculty of Electrical Engineering, University of Kassel, Germany
62. Kato N, Kurozumi K, Susuki N, Muroyama S (2001) Hybrid power supply system composed of photovoltaic and fuel cell systems. In: Conférence INTELEC, pp 631–635
63. Mokadem ELM, Nichita C, Barkat G, Dakyo B (2002) Control strategy for stand alone wind-diesel hybrid system using a speed model. In: 7th international ELECTRIMACS congress, Montréal
64. Jurado F, Saenz R (2002) Neuro-fuzzy control for autonomous wind-diesel systems using biomass. Renew Energy 27:39–56
65. Bekka H, Taraft S, Rekioua D, Bacha S (2013) Power control of a wind generator connected to the grid in front of strong winds. J Electr Sys 9(3):267–278
66. Gevorgian V, Touryan K, Bezrukikh P, Karghiev V, Bezrukikh P (1999) Wind-diesel hybrid systems for Russia's Northern territories, presented at Windpower '99, Burlington, Vermont, Jun 20–23 1999
67. Manwell JF, Mcgowan JG, Abdulwahid U (2000) Simplified performance model for hybrid wind diesel systems, renewable energy: the energy for the 21st century. In World renewable energy congress N°6, Brighton, ROYAUME-UNI, pp 1183–1188
68. Lysen EH (2000) Hybrid technology. http://climatetechwiki.org/technology/hybrid-technology
69. Prasad AR, Natarajan E (2006) Optimization of integrated photovoltaic-wind power generation systems with battery storage. Energy 31:1943–1954
70. Diaf S, Haddadi M, Belhamel M (2006) Techno economic analysis of a hybrid system (photovoltaic/wind) independent website for the Adrar. Renew Energy J 9(3):127–134
71. Dali M, Belhadj J, Roboam X, Blaquiere JM (2007) Control and energy management of a wind-photovoltaic hybrid system. In: Proceedings of 12th European conference on power electronics and applications (EPE'2007), Aalborg, Denmark, 02–05 Sep 2007, pp 1–10

72. Razak JA, Sopian K, Ali Y, Alghoul MA, Zaharim A, Ahmad I (2009) Optimization of PV-wind-hydro-diesel hybrid system by minimizing excess capacity. Eur J Sci Res 25(4):663–671
73. Idjdarene K, Rekioua D, Rekioua T, Tounzi A (2009) Direct torque control strategy for a variable speed wind energy conversion system associated to a flywheel energy storage system. In: Proceedings—international conference on developments in eSystems engineering, DeSE 2009:17–22
74. Rekioua D, Bensmail S, Bettar N (2013) Development of hybrid photovoltaic-fuel cell system for stand-alone application. Int J Hydrogen Energy 39(3):1604–1611
75. Lalouni S, Rekioua D, (2009) Modeling and simulation of a photovoltaic system using fuzzy logic controller. Proceedings—international conference on developments in eSystems engineering, DeSE 2009:23–28
76. http://www.windsystemsmag.com/article/detail/53/maintenance
77. http://www.windustry.org/resources/how-much-do-wind-turbines-cost
78. Odo FC, Offiah SU, Ugwuoke PE (2012) Weibull distribution-based model for prediction of wind potential in Enugu, Nigeria. Adv Appl Sci Res 3(2):1202–1208
79. Cámara EM, Macías EJ (2010) InTech book chapter 2. Wind Power Book
80. Muyeen SM, Al-Durra A, Tamura J (2011) Energy conversion and management 52:2688–2694

Chapter 2
Wind Energy Conversion and Power Electronics Modeling

2.1 Wind Energy Conversion Modeling

The global structure of a wind energy conversion system consists generally of the aerogenerator, the power electronics, and the load (Fig. 2.1). The modelization of this system is presented as follows. Each component can be modeled separately.

2.1.1 Aerogenerator Modeling

2.1.1.1 Velocity Modeling

We have defined in Chap. 1 that there are three important values of wind speeds which are the cut-in wind speed, the nominal wind speed, or rated wind speed, and finally the cut out wind speed. These speeds define the wind turbine operation, and it is commonly measured with an anemometer or by a weather station in (m/s) or (km/h).

The usual models are [1]:

- Weibull distribution,
- Weibull distribution hybrid,
- Rayleigh distribution.

Weibull Distribution

This is the general model that describes the wind speed variations. This model optimizes the turbine design to minimize the electricity production costs. The Weibull coefficient reflects the distribution of wind speeds, and it is determined by the Weibull distribution curve. The function of probability density Weibull is given as [1]:

D. Rekioua, *Wind Power Electric Systems*, Green Energy and Technology,
DOI: 10.1007/978-1-4471-6425-8_2,

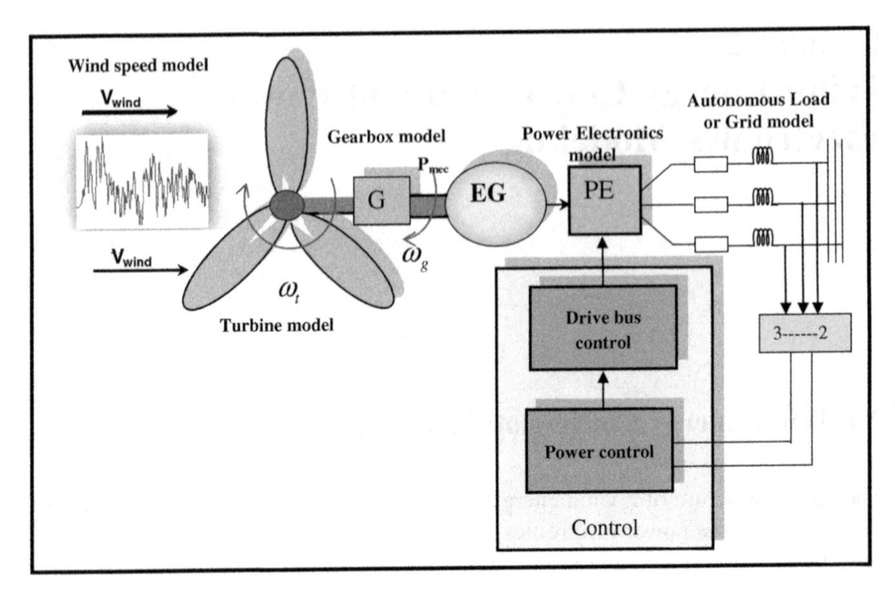

Fig. 2.1 Structure of a conversion wind system modeling

$$f(v) = k \cdot c^{-k} \cdot v^{k-1} e^{-(v/c)^k} \tag{2.1}$$

where the probability density $f\,(V)$ is the frequency distribution of the measured velocities, k and c are Weibull parameters, where the parameter k (dimensionless shape factor) characterizes the shape of the frequency distribution, while C (scale factor). Determines the quality of the wind this one has the same unit as the speed. These two parameters k and C are used for the calculation of mean wind speed. We use the following expression to obtain the scale factor [1]:

$$C = \frac{v_{\text{mean}}}{\gamma \cdot \left(1 + \frac{1}{k}\right)} \tag{2.2}$$

where γ is the gamma function.

The area under the curve is defined by [1]:

$$f(v) = 1 - e^{-(v/c)^k}. \tag{2.3}$$

Weibull Distribution Hybrid

Hybrid Weibull distribution is used where the calm frequency recorded higher or equal to 15 %. Generally, this proportion cannot be neglected and must be taken into account during the site characterization in terms of wind. This distribution is as follows:

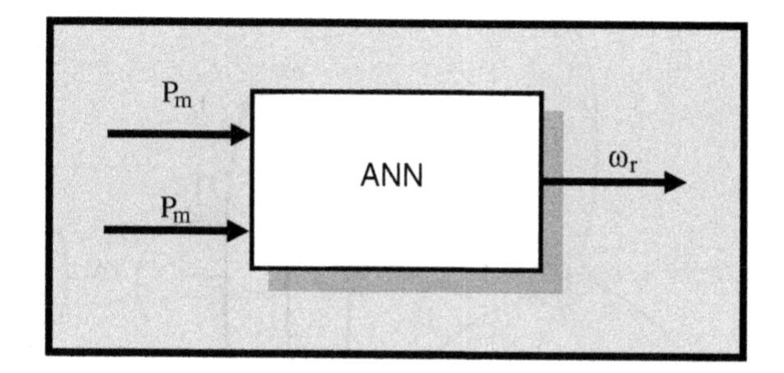

Fig. 2.2 Estimated of wind speed using neural networks

$$f(v) = (1 - ff_0) \cdot \left(\frac{k}{c}\right) \cdot \left(\frac{v}{c}\right)^{k-1} e^{-\left(\frac{v}{c}\right)^k} \quad \text{for } v > 0 \tag{2.4}$$

$$f(v) = ff_0 \quad \text{for } v < 0 \tag{2.5}$$

where ff_o is the calm frequency.

Rayleigh Distribution

The Rayleigh distribution is a special case of the Weibull distribution when the shape factor k is equal to 2, its probability density is given by:

$$f(v) = 2 \cdot \frac{v}{c^2} \cdot e^{-(v/c)^2} \tag{2.6}$$

Wind Speed Estimation

The wind speed can be estimated using neural networks. The inputs to the ANN are the rotor speed ω_r and mechanical power P_m. It is obtained using the relation (Fig. 2.2).

$$P_m = \omega_r \cdot \left(J \frac{d\omega_r}{dt}\right) + P_e. \tag{2.7}$$

2.1.1.2 Shaft Model

Two-Mass Model

The two-mass model is very much used in scientific researches [2–5]. It is represented in Fig. 2.3.

Driving by the aerodynamic torque T_a, the rotor of the wind turbine runs at the speed ω_r. The low-speed shaft torque T_{ls} acts as a braking torque on the rotor (Fig. 2.3). The dynamics of the rotor is characterized by the first-order differential equation:

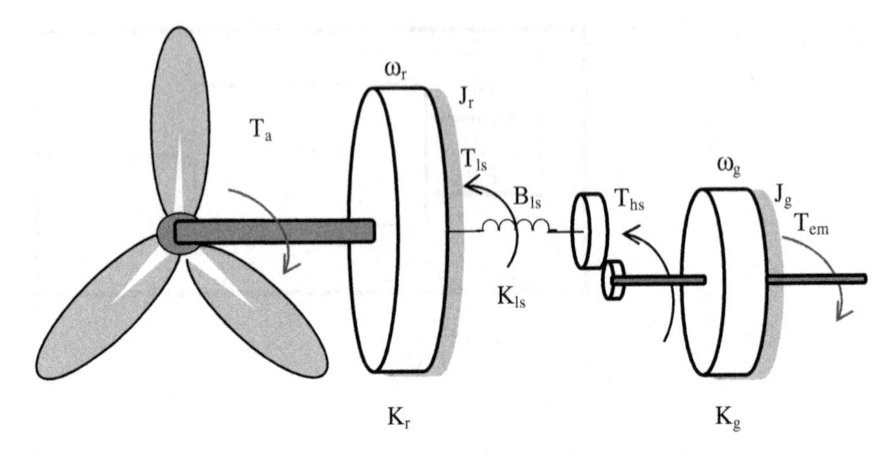

Fig. 2.3 Drive train dynamics

$$J_r \dot{\omega}_t = T_a - T_{\text{ls}} - K_r \omega_t \tag{2.8}$$

The low-speed shaft results T_{ls} from the torsion and friction effects due to the difference between ω_t and the low-shaft speed ω_{ls}. This torque act as a breaking torque on the rotor.

$$T_{\text{ls}} = B_{\text{ls}}(\theta_t - \theta_{\text{ls}}) + K_{\text{ls}}(\omega_t - \omega_{\text{ls}}) \tag{2.9}$$

Using the gearbox G, torque and speed of low shaft are provided to generate a torque on high shaft,

$$T_{\text{hs}} = \frac{T_{\text{ls}}}{G} \tag{2.10}$$

And we have

$$\begin{aligned} \theta_g &= G \cdot \theta_{\text{ls}} \\ \omega_g &= G \cdot \omega_{\text{ls}} \end{aligned} \tag{2.11}$$

Practically, we have

$$T_{\text{hs}} = G \cdot \frac{T_{\text{ls}}}{G} \tag{2.12}$$

Through the gearbox, the low-shaft speed ω_{ls} is increased by the gearbox ratio to obtain the generator speed ω_g, while the low-speed shaft torque T_{ls} is increased. If we assume an ideal gearbox with a ratio G, we can write the following:

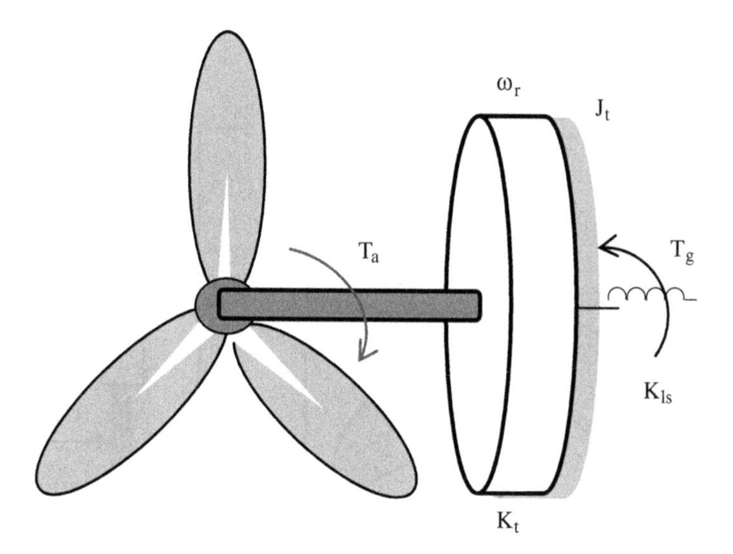

Fig. 2.4 One-mass model

$$G = \frac{T_{\text{ls}}}{T_{\text{hs}}} = \frac{\omega_g}{\omega_{\text{ls}}} = \frac{\theta_g}{\theta_{\text{ls}}} \tag{2.13}$$

The generator is driven by the high-speed shaft Torque T_{hs} and braked by the generator electromagnetic torque T_{em}. Its dynamic is given as:

$$J_g \dot{\omega}_g = T_{\text{hs}} - K_g \omega_g - T_{\text{em}} \tag{2.14}$$

One-Mass Model

The rigidity of the main shaft of the drive of great wind turbines can be considered perfect. Also, two-mass model can then be reduced to one-mass model. The one-mass model comprises a single inertia and a single friction coefficient with all external friction coefficients (Fig. 2.4).

This model assumes that:

- Drive shafts are perfectly rigid,
- The speed gearbox is ideal,
- The generator inertia can be neglected compared with the turbine one or reduced to the low shaft.

With this model, the drive flexibility is not taken into account thus we will have the possibility of neglecting some mechanical coupling properties [4, 5].

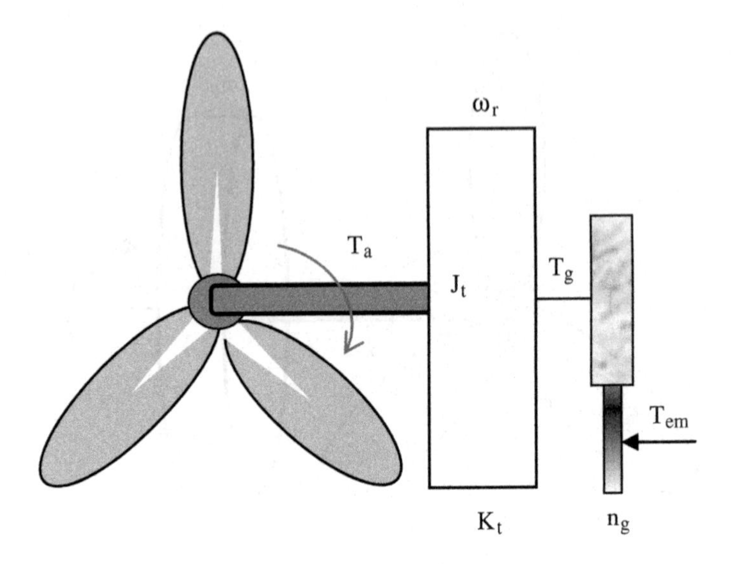

Fig. 2.5 One-mass model reduced to the low shaft

One-Mass Model Reduced to the Low Shaft

The turbine inertia model in this case can be represented as in Fig. 2.5.

Since the low shaft is infinitely rigid (torsion coefficient on the low shaft B_{ls} is infinite), then:

$$\theta_t = \theta_{\mathrm{ls}} \quad \text{and} \quad \omega_t = \omega_{\mathrm{ls}}, \tag{2.15}$$

We obtain

$$G = \frac{T_{\mathrm{ls}}}{T_{\mathrm{hs}}} = \frac{\omega_g}{\omega_t} = \frac{\theta_g}{\theta_t} \tag{2.16}$$

$$\begin{aligned} J_r \cdot \dot{\omega}_t &= T_a - G \cdot T_{\mathrm{hs}} - K_r \cdot \omega \\ G \cdot J_g \cdot \dot{\omega}_t &= T_{\mathrm{hs}} - K_g \cdot G \cdot \omega_t - T_{\mathrm{em}} \end{aligned} \tag{2.17}$$

Multiplying the second expression of Eq. 2.17 and by adding the two equations, we obtain as follows:

$$J_t \dot{\omega}_t = T_a - K_t \omega_t - T_g \tag{2.18}$$

with

$$J_t = J_r + G^2 J_g \tag{2.19}$$

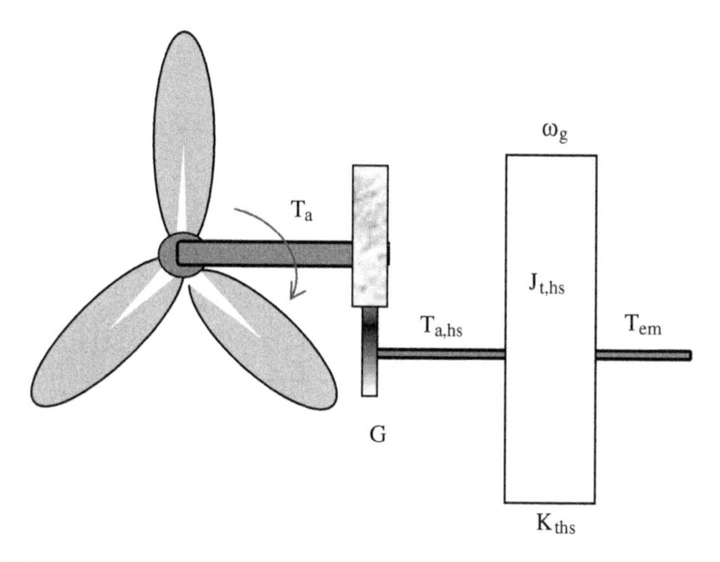

Fig. 2.6 One-mass model reduced to high-speed shaft

$$K_t = K_r + G^2 K_g \tag{2.20}$$

$$T_g = G \cdot T_{\text{em}} \tag{2.21}$$

where J_t, K_t, and T_g are, respectively, the inertia, the friction coefficient of the turbine, and the electromagnetic torque of the generator reduced to the low shaft.

The generator inertia reduced to the low shaft is often neglected next to the rotor inertia. The one-mass model reduced to the low shaft is generally used in the control of aerogenerator.

One-Mass Model Reduced to High-Speed Shaft

The proposed mechanical model consists of the total turbine inertia reduced to the generator shaft (high-speed shaft) [5]. This model is illustrated in Fig. 2.6.

We use the same assumptions for the model reduced to low shaft, the equations became:

$$\begin{aligned} \frac{J_r}{G}\dot{\omega}_g &= T_a - GT_{\text{ls}} - \frac{K_r}{G}\omega_g \\ J_g\dot{\omega}_g &= T_{\text{hs}} - K_g\omega_g - T_{\text{em}} \end{aligned} \tag{2.22}$$

Dividing the second expression of Eq. 2.19 by G and summing the two expressions, we obtain

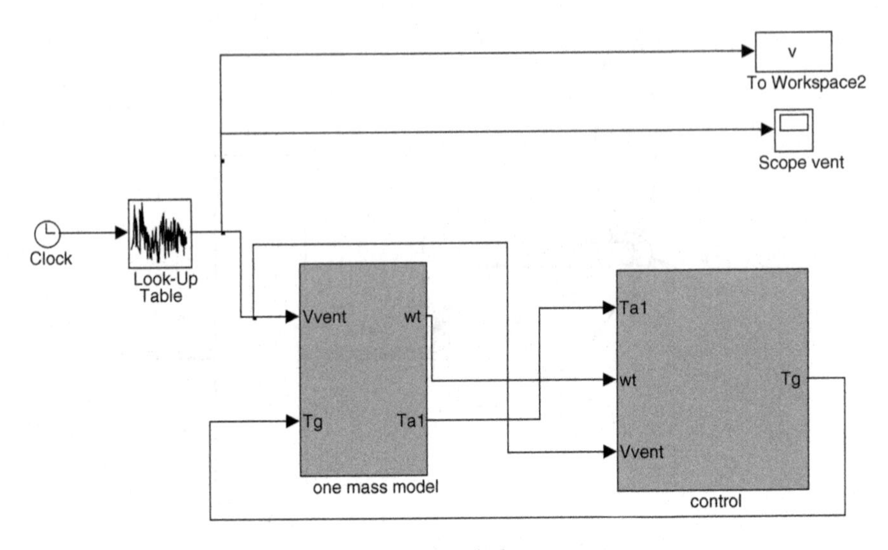

Fig. 2.7 One-mass model reduced to high-speed shaft

$$J_{t_{\mathrm{hs}}}\dot{\omega}_g = T_{a_{\mathrm{hs}}} - K_{t_{\mathrm{hs}}}\omega_g - T_{\mathrm{em}} \tag{2.23}$$

with

$$J_{t_{\mathrm{hs}}} = J_g + \frac{J_r}{G^2} \tag{2.24}$$

$$K_{t_{\mathrm{hs}}} = K_g + \frac{K_r}{G^2} \tag{2.25}$$

$$T_{a_{\mathrm{hs}}} = \frac{T_a}{G} \tag{2.26}$$

where $J_{t_{\mathrm{hs}}}$, $K_{t_{\mathrm{hs}}}$, and $T_{a_{\mathrm{hs}}}$ are, respectively, the inertia, the external friction coefficient of the turbine, and aerodynamic torque of the generator reduced the high shaft.

It should be noted that the generator inertia is very small compared with the turbine inertia reported on this shaft. The one-mass model reduced to the high shaft is generally used in the generator control. The one-mass model under MATLAB/Simulink is shown in Fig. 2.7.

Dynamic Turbine Model

The diagram of the dynamic model of the wind-based turbine is given in Fig. 2.8.

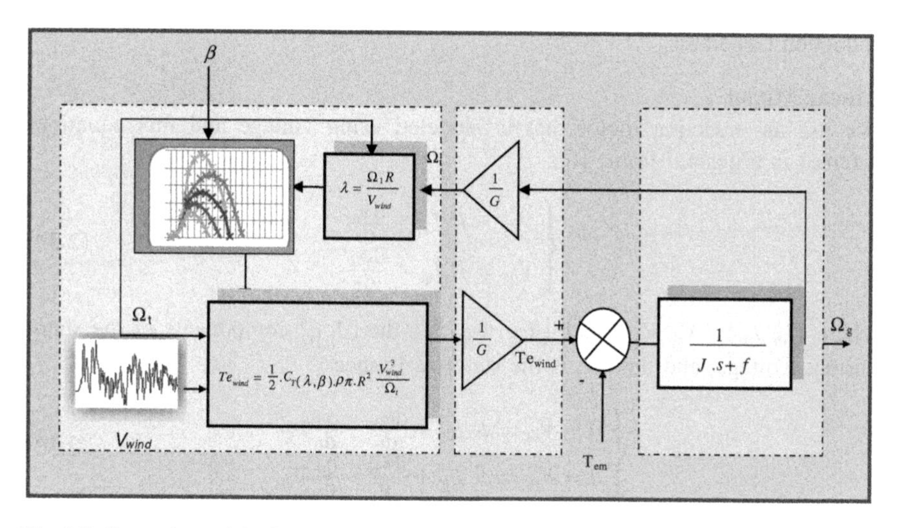

Fig. 2.8 Dynamic model of the wind turbine

2.1.1.3 Electrical Generators Modeling

Permanent Magnet Synchronous Generators

Due to absence of the field current and field winding, permanent magnet machines exhibit high efficiency in operation, simple and robust structure in construction and high power-to-weight ratio. The attractiveness of the permanent magnet machines is further enhanced by the availability of high-energy rare-earth permanent magnet materials such as SmCo and NdFeB. However, the speed control of permanent magnet DC motor via changing the field current is not possible.

The voltage equations are given by:

$$\begin{bmatrix} v_a \\ v_b \\ v_c \end{bmatrix} = R_s \begin{bmatrix} i_a \\ i_b \\ i_c \end{bmatrix} + L_c \frac{\mathrm{d}}{\mathrm{d}t} \begin{bmatrix} i_a \\ i_b \\ i_c \end{bmatrix} + \begin{bmatrix} E_a \\ E_b \\ E_c \end{bmatrix} \tag{2.27}$$

where $v_{a,\ b,\ c}$ the machine phase voltages, $i_{a,\ b,\ c}$ the current phase voltages, L_c the cyclic inductance, R_s resistance of the stator winding, $E_{a,\ b,\ c}$ represents the induced E.F.M (electric force motor) in the stator phase windings.

The electromagnetic torque is expressed as:

$$T_{\mathrm{em}} = \frac{P}{\omega}(E_a \cdot i_a + E_b \cdot i_b + E_c \cdot i_c) \tag{2.28}$$

where P the number of pole pairs, ω the rotor speed $(\omega = P \cdot \Omega)$.

Induction Generator

Linear Model

We use an induction motor that is modeled using voltage and flux equations referred in a general frame [6]:

$$\begin{cases} V_{ds} = R_s I_{ds} + \frac{d\Phi_{ds}}{d_t} \\ V_{qs} = R_s I_{qs} + \frac{d\Phi_{qs}}{d_t} \end{cases} \tag{2.29}$$

where (I_{ds}, I_{qs}), (V_{ds}, V_{qs}) and (Φ_{ds}, Φ_{qs}) are the (d, q) components of the stator current, voltage, and flux, R_s is the stator resistance.

$$\begin{cases} 0 = V_{dr} = R_r I_{dr} + \frac{d\Phi_{dr}}{d_t} + \frac{d\theta}{d_t}\Phi_{qr} \\ 0 = V_{qr} = R_r I_{qr} + \frac{d\Phi_{qr}}{d_t} - \frac{d\theta}{d_t}\Phi_{dr} \end{cases} \tag{2.30}$$

where I_{dr} and I_{qr} are (d, q) rotor current, R_r is the rotor resistance, and Φ_{dr} and Φ_{qr} are (d, q) rotor flux.

We have

$$\begin{bmatrix} \Phi_{ds} \\ \Phi_{qs} \\ \Phi_{rd} \\ \Phi_{qr} \end{bmatrix} = \begin{bmatrix} L_s & 0 & L_m & 0 \\ 0 & L_s & 0 & L_m \\ L_m & 0 & L_r & 0 \\ 0 & L_m & 0 & L_r \end{bmatrix} \cdot \begin{bmatrix} i_{ds} \\ i_{qs} \\ i_{dr} \\ i_{qr} \end{bmatrix}$$

We obtain the following mathematical model:

$$\begin{bmatrix} \frac{di_{ds}}{d_t} \\ \frac{di_{qs}}{d_t} \\ \frac{di_{dr}}{d_t} \\ \frac{di_{qr}}{d_t} \end{bmatrix} = \frac{1}{\sigma}\begin{bmatrix} -\frac{R_s}{L_s} & \frac{P\cdot\omega_r\cdot L_m^2}{L_s\cdot L_r} & \frac{L_m\cdot R_r}{L_s\cdot L_r} & \frac{P\cdot\omega_r\cdot L_m}{L_s} \\ -\frac{P\cdot\omega_r\cdot L_m^2}{L_s\cdot L_r} & -\frac{R_s}{L_s} & \frac{P\cdot\omega_r\cdot L_m}{L_s} & \frac{L_m\cdot R_r}{L_s\cdot L_r} \\ \frac{L_m\cdot R_s}{L_s\cdot L_r} & -\frac{P\cdot\omega_r\cdot L_m}{L_r} & -\frac{R_r}{L_r} & -P\cdot\omega_r \\ \frac{P\cdot\omega_r\cdot L_m}{L_r} & \frac{L_m\cdot R_s}{L_s\cdot L_r} & P\cdot\omega_r & -\frac{R_r}{L_r} \end{bmatrix}\begin{bmatrix} i_{ds} \\ i_{qs} \\ i_{dr} \\ i_{qr} \end{bmatrix} + \frac{1}{\sigma}\begin{bmatrix} \frac{1}{L_s} & 0 \\ 0 & \frac{1}{L_s} \\ -\frac{L_m}{L_s\cdot L_r} & 0 \\ 0 & -\frac{L_m}{L_s\cdot L_r} \end{bmatrix}\begin{bmatrix} v_{ds} \\ v_{qs} \end{bmatrix} \tag{2.31}$$

with σ is the leakage coefficient.

Mechanical equation:

$$T_{em} - T_{Load} = J_{mot} \cdot \frac{d\omega_r}{dt} \tag{2.32}$$

with J_{mot} the inertia of the AC motor.

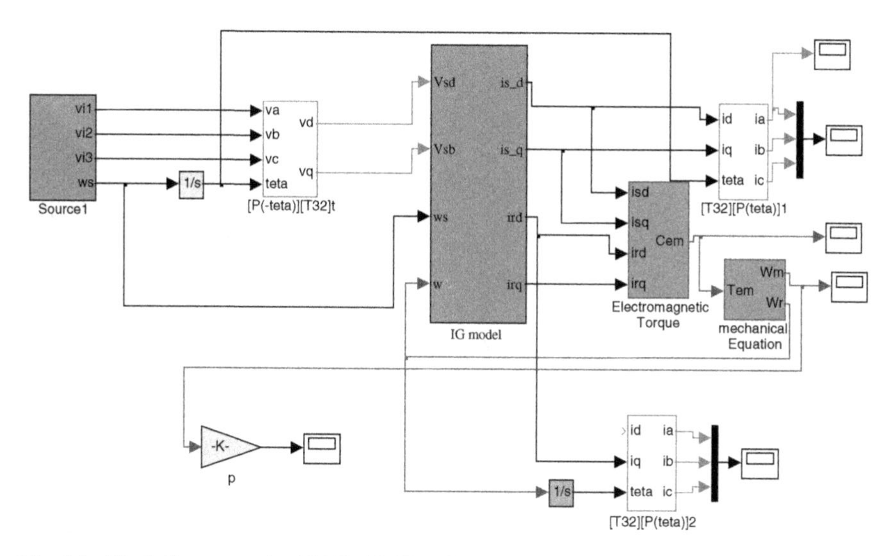

Fig. 2.9 *Block* diagram under MATLAB/Simulink of induction generator in linear operation

The electromagnetic torque can be written as:

$$T_{\text{em}} = P \cdot \left(\phi_{\text{ds}} \cdot i_{\text{qs}} - \phi_{\text{qs}} \cdot i_{\text{ds}}\right) \tag{2.33}$$

Figure 2.9 shows the block diagram under MATLAB/Simulink of induction generator in linear operation.

Some simulation results are presented in Fig. (2.10).

Saturated Model

The linear model of the induction machine is widely known and used. It yields results relatively accurate when the operating point studied is not so far from the conditions of the model parameter identification. This is often the case when the motor operating, at rated voltage, is studied. As the air gap of induction machines is generally narrow, the saturation effect is not negligible in this structure. So, to improve the accuracy of simulation studies, especially when the voltage is variable, the nonlinearity of the iron has to be taken into account in the machine model. This becomes a necessary condition to study an autonomous induction generator because the linear model is not able to describe the behavior of the system. Thus, only approaches, which take account of the saturation effect, can be used. This effect is not easy to yield with using three-phase classical models. So, we usually adopt diphase approaches to take globally account of the magnetic nonlinearity. This evidently supposes some simplifying hypotheses. Indeed, the induction is considered homogenous in the whole structure. Moreover, the use of diphase model supposes that the saturation effect is considered only on the first

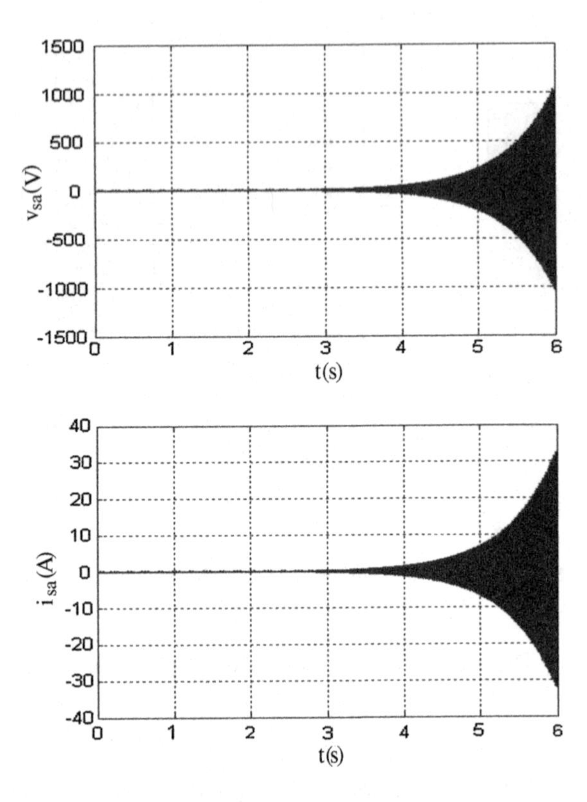

Fig. 2.10 Stator current and voltage of induction generator in linear operation

harmonics and does not affect the sinusoidal behavior of the variables. We adopt the diphase model of the induction machine expressed in the stator frame. The classical electrical equations are written as follows:

$$\begin{bmatrix} V_{\mathrm{ds}} \\ V_{\mathrm{qs}} \\ 0 \\ 0 \end{bmatrix} = \begin{bmatrix} R_s & -\omega_s \cdot l_s & 0 & -\omega_s \cdot L_m \\ \omega_s \cdot l_s & R_s & \omega_s \cdot L_m & 0 \\ -R_r & \omega_r \cdot l_r & R_r & -\omega_r \cdot (l_r + L_m) \\ -\omega_r \cdot l_r & -R_r & \omega_r \cdot (l_r + L_m) & R_r \end{bmatrix} \begin{bmatrix} i_{\mathrm{ds}} \\ i_{\mathrm{qs}} \\ i_{\mathrm{d0}} \\ i_{\mathrm{q0}} \end{bmatrix}$$
$$+ \begin{bmatrix} l_s & 0 & L_m + L'_m \cdot \frac{i_{\mathrm{d0}}^2}{|i_1|} & L'_m \cdot \frac{i_{\mathrm{d0}} . i_{\mathrm{q0}}}{|i_m|} \\ 0 & l_s & L'_m \cdot \frac{i_{\mathrm{d0}} . i_{\mathrm{q0}}}{|i_0|} & L_m + L'_m \cdot \frac{i_{\mathrm{q0}}^2}{|i_0|} \\ -l_r & 0 & l_r + L_m + L'_m \cdot \frac{i_{\mathrm{d0}}^2}{|i_0|} & L'_m \cdot \frac{i_{\mathrm{d0}} . i_{\mathrm{q0}}}{|i_0|} \\ 0 & -l_r & L'_m \cdot \frac{i_{\mathrm{d0}} \cdot i_{\mathrm{q0}}}{|i_0|} & l_r + L_m + L'_m \cdot \frac{i_{\mathrm{q0}}^2}{|i_0|} \end{bmatrix} \cdot \begin{bmatrix} \frac{\mathrm{d}i_{\mathrm{ds}}}{\mathrm{d}t} \\ \frac{\mathrm{d}i_{\mathrm{qs}}}{\mathrm{d}t} \\ \frac{\mathrm{d}i_{\mathrm{d0}}}{\mathrm{d}t} \\ \frac{\mathrm{d}i_{\mathrm{q0}}}{\mathrm{d}t} \end{bmatrix} \tag{2.34}$$

where R_s, l_s, R_r, and l_r are the stator and rotor phase resistances and leakage inductances, respectively, L_m is the magnetizing inductance and $\omega = p \cdot \Omega$.

Besides, V_{ds}, i_{ds}, V_{qs}, and i_{qs} are the d-q stator voltages and currents, respectively. i_{d0} and i_{q0} are the magnetizing currents, along the d and q axis, given by:

$$\begin{cases} i_{d0} = i_{ds} + i_{dr} \\ i_{q0} = i_{qs} + i_{qr} \end{cases} \quad (2.35)$$

where i_{dr} and i_{qr} are the d–q rotor currents.

Thus, the saturation effect is taken into account by the expression of the magnetizing inductance L_m with respect to the magnetizing current i_0 defined as:

$$i_0 = \sqrt{i_{d0}^2 + i_{q0}^2} \quad (2.36)$$

To express L_m in function of i_0, we can use different approaches in this case, we use a polynomial approximation of degree 12 [6].

$$\begin{cases} L_m = f(|i_0|) = \sum_{j=0}^{n} a_j \,.\, |i_0|_0^j \\ L'_m = \frac{dL_m}{d|i_0|} = \frac{d}{d|i_0|} f(|i_0|) = \sum_{j=0}^{n} j \,.\, a_j \,.\, |i_0|_0^{j-1} \end{cases} \quad (2.37)$$

Figure 2.11 shows the block diagram under MATLAB/Simulink of induction generator in saturated operation.

Some simulation results are presented Fig. (2.12).

Double-Fed Induction Generator

The DFIG dynamic model in Park's reference is expressed as follows [7]. Voltage equations are given by:

$$\begin{cases} v_{ds} = R_s i_{ds} + \frac{d}{dt}\phi_{ds} - \omega_s \phi_{qs} \\ v_{qs} = R_s i_{qs} + \frac{d}{dt}\phi_{qs} + \omega_s \phi_{ds} \\ v_{dr} = R_r i_{dr} + \frac{d}{dt}\phi_{dr} - (\omega_s - \omega)\phi_{qr} \\ v_{qr} = R_r i_{qr} + \frac{d}{dt}\phi_{qr} + (\omega_s - \omega)\phi_{dr} \end{cases} \quad (2.38)$$

Flux linkage equations are obtained from:

$$\begin{cases} \phi_{ds} = L_s i_{ds} + L_m i_{dr} \\ \phi_{qs} = L_s i_{qs} + L_m i_{qr} \\ \phi_{dr} = L_r i_{dr} + L_m i_{ds} \\ \phi_{qr} = L_s i_{qr} + L_m i_{qs} \end{cases} \quad (2.39)$$

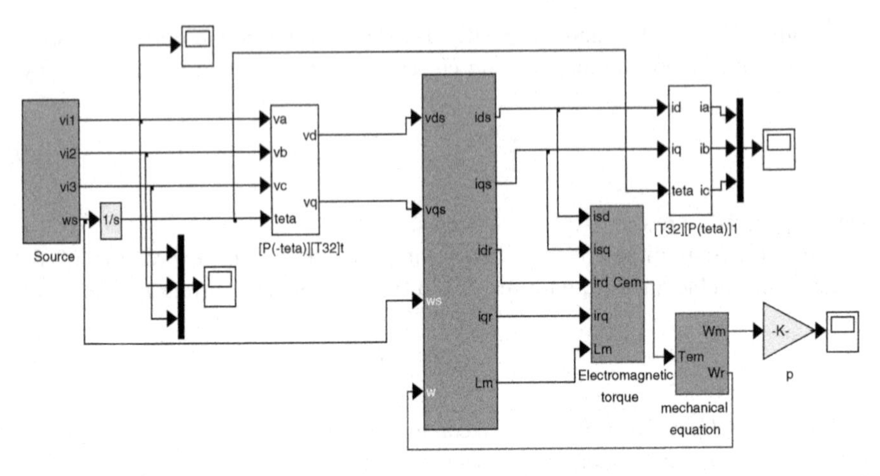

Fig. 2.11 *Block* diagram under MATLAB/Simulink of induction generator in saturated operation

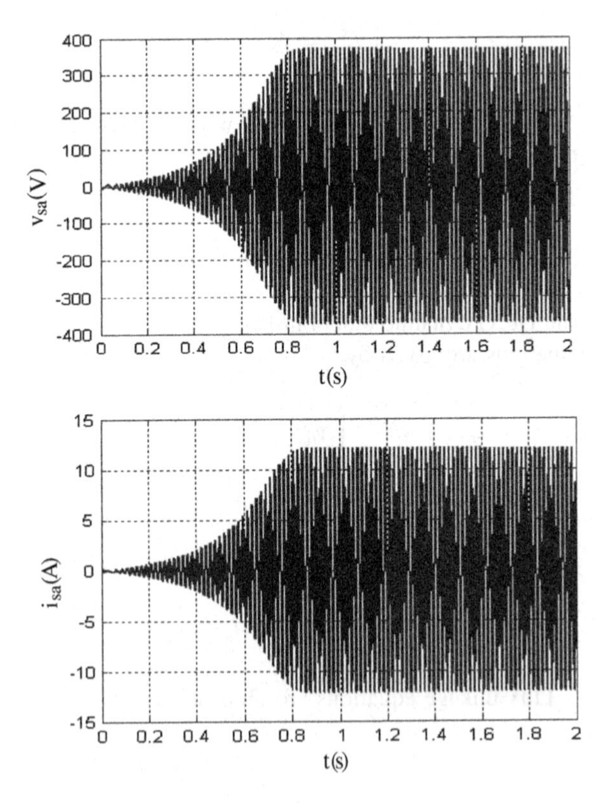

Fig. 2.12 Stator current and voltage of induction generator in saturated operation

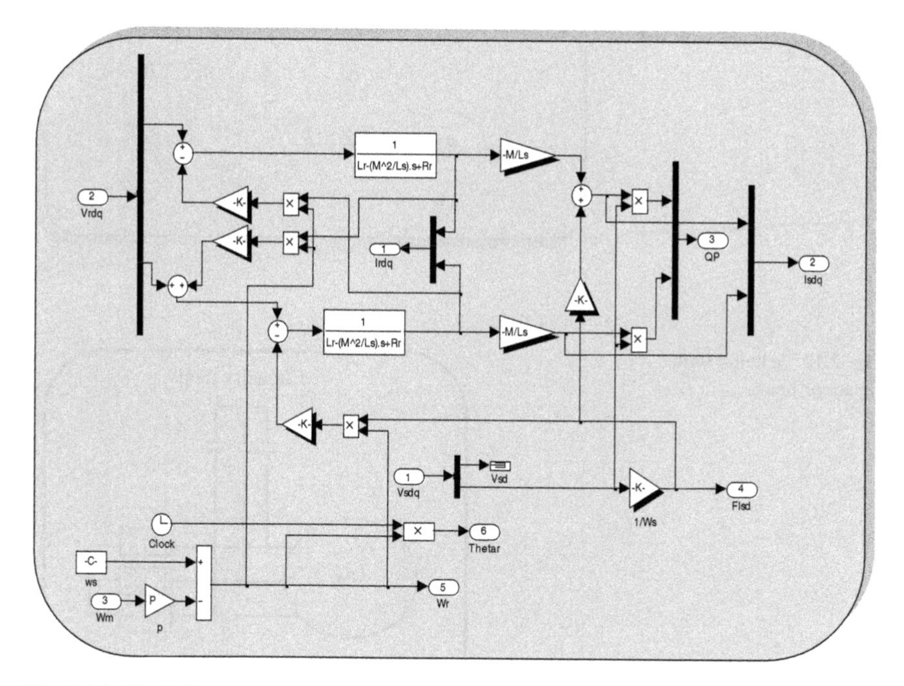

Fig. 2.13 *Block* diagram under MATLAB/Simulink of DFIG

Electromagnetic torque equation is

$$T_{\mathrm{em}} = p\left(\phi_{\mathrm{ds}} \cdot i_{\mathrm{qs}} - \phi_{\mathrm{qs}} \cdot i_{\mathrm{ds}}\right) \tag{2.40}$$

Figure 2.13 shows the block diagram under MATLAB/Simulink of DFIG.

2.2 Power Electronics Modeling

2.2.1 Soft Starter

Generally, it is used to start induction generator. It is an AC–AC direct converter with two thyristors back-to-back in each phase (see Fig. (2.14)).

2.2.2 Capacitor Bank

A capacitor bank is generally used in induction generators to obtain the magnetizing flux. It is connected to the generator in autonomous or stand-alone system (see Fig. (2.15)).

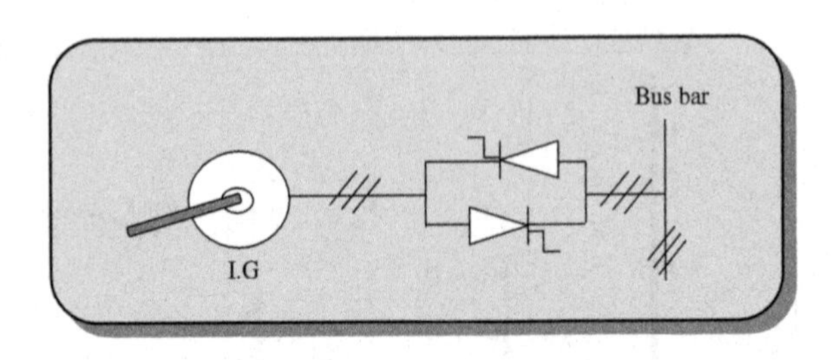

Fig. 2.14 Scheme with soft starter

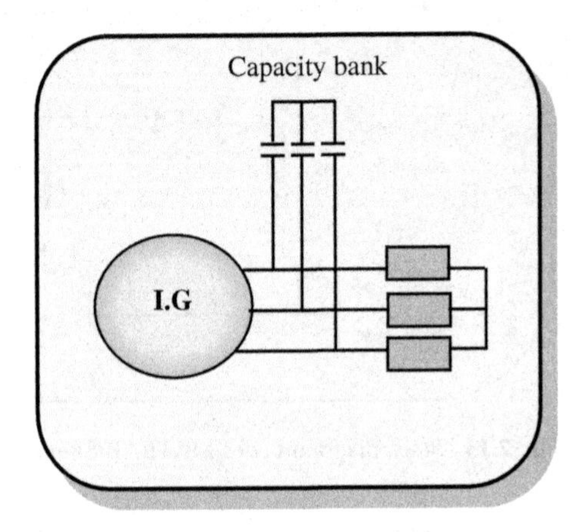

Fig. 2.15 Scheme with capacitor bank

2.2.3 Diode Rectifier

This topology is simple, low cost, and there is less harmonic distortion and low losses. Low efficiency is its most disadvantage (see Fig. (2.16)).

2.2.4 The Back-to-Back PWM-VSI

The back-to-back PWM-VSI is a bidirectional power converter consisting of two conventional pulse width modulated (PWM) voltage source converters (VSC), as shown in Fig 2.17. One of the converters operates in the rectifying mode, while the other converter operates in the inverting mode. These two converters are connected together via a DC-link consisting of a capacitor.

This topology provides active and reactive control, high-power factor in the generator side and high efficiency, but its disadvantages are short life, high-switching losses, and frequency harmonics.

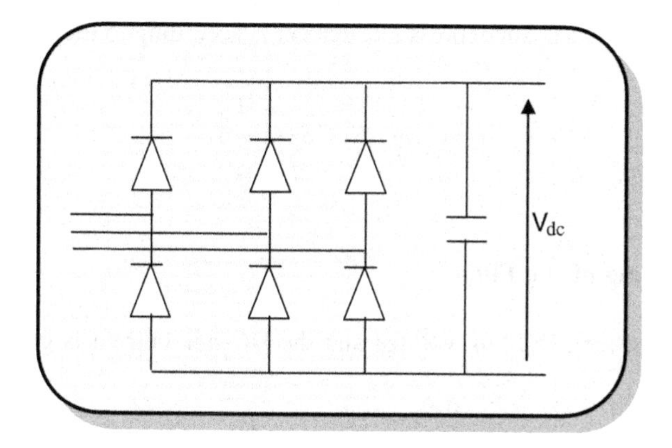

Fig. 2.16 Scheme with diode rectifier

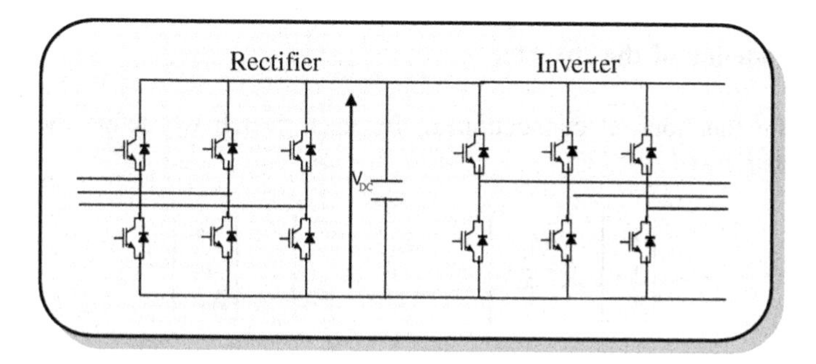

Fig. 2.17 The back-to-back PWM-VSI converter

2.2.4.1 Modeling of the Rectifier

Using the functions of connection S_i for each switch, we define the rectifier transfer matrix as [7]:

$$\begin{bmatrix} V_{\mathrm{DC}}^{+} \\ V_{\mathrm{DC}}^{-} \end{bmatrix} = \begin{bmatrix} S_1 & S_3 & S_5 \\ S_2 & S_4 & S_6 \end{bmatrix} \cdot \begin{bmatrix} v_a \\ v_b \\ v_c \end{bmatrix} \tag{2.41}$$

with S_i, $i = 1$–6, function of connection associated with each switch of the rectifier. The DC-link voltage is obtained as

$$V_{\mathrm{DC}} = V_{\mathrm{DC}}^{+} - V_{\mathrm{DC}}^{-}. \tag{2.42}$$

In the same way, we can express DC current i_s according to the currents among

$$i_s = \begin{bmatrix} S_1 & S_3 & S_5 \end{bmatrix} \begin{bmatrix} i_1 \\ i_2 \\ i_3 \end{bmatrix}. \tag{2.43}$$

2.2.4.2 Modeling of the Filter

The relation between DC-link voltage and the DC-link current is given as [7]:

$$\frac{\mathrm{d}V_{\mathrm{dc}}}{\mathrm{d}t} = \frac{1}{C_{\mathrm{DC}}}(i_s - i_0) \tag{2.44}$$

where C_{DC} is the DC-link capacitance.

2.2.4.3 Modeling of the Inverter

We use the functions of connection S_i, for each switch we define the inverter transfer matrix as [7]:

$$\begin{bmatrix} v_a \\ v_b \\ v_c \end{bmatrix} = \frac{V_{\mathrm{DC}}}{3} \begin{bmatrix} 2 & -1 & -1 \\ -1 & 2 & -1 \\ -1 & -1 & 2 \end{bmatrix} \cdot \begin{bmatrix} S_7 \\ S_9 \\ S_{11} \end{bmatrix} \tag{2.45}$$

with S_i, $i = 7\text{–}12$, function of connection associated with each switch of the inverter.

For the two converters, the technique of control by pulse width the modulation (PWM) is applied as well as the complementary order for each arm is adopted.

2.2.5 *Tandem Converter*

It consists of two parallel inverters. The first one operates as a current source converter (CSC), and the second operates as a back-to-back PWM voltage source inverter (VSI) Fig. (2.18).

This structure has several advantages [8].

- Low-switching frequency of the primary converter.
- Low level of the switched current of the secondary converter.

2.2.6 *Matrix Converter*

The simplified three phases matrix converter topology incorporated in system of wind turbine is shown in Fig. 2.19. Matrix converter consists of nine bidirectional

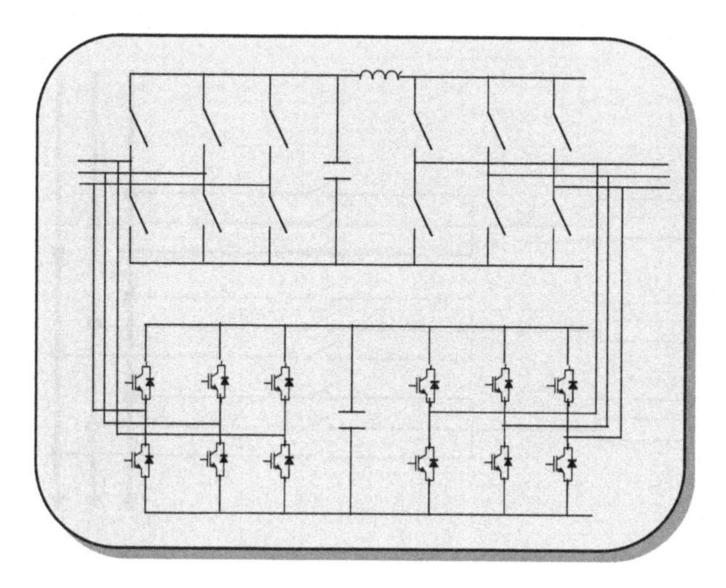

Fig. 2.18 Tandem converter topology

switches, and each output phase is associated with three switches set connected to three input phases. This configuration of bidirectional switch enables to connect any input phase *a, b,* or *c* to any output phase *A, B,* or *C* at any instant. Since the matrix converter is supplied by the voltage source, the input phases must not be shorted, and due to the inductive nature of the load, the output phases must not be open. If the switch function of a switch S_{ij} in Fig. 2.19 is defined as [1, 2].

$$S_{ij}(t) = \begin{cases} 1 & S_{ij} \text{ is closed} \\ 0 & S_{ij} \text{ is open} \end{cases} \quad i \in \{a, b, c\},\ j \in \{A, B, C\} \tag{2.46}$$

The following constraints are applied to the switching functions:

$$S_{\mathrm{aj}} + S_{\mathrm{bj}} + S_{\mathrm{cj}} = 1, \quad j \in \{A, B, C\} \tag{2.47}$$

This constraint comes from the fact that MC is supplied from a voltage source and generally feeds an inductive load. Connecting more than one input phases to one output phase causes a short circuit between two input sources. Also, disconnecting all input phases from an output phase results an open circuit in the load [3].

The aim of the modulation strategy is to synthesize the output voltages from the input voltages and the input currents from the output currents.

The three-phase matrix converter can be represented by a set of three matrixes from because the nine bidirectional switches can connect one input phase to one output phase directly without any intermediate energy storage elements. Therefore,

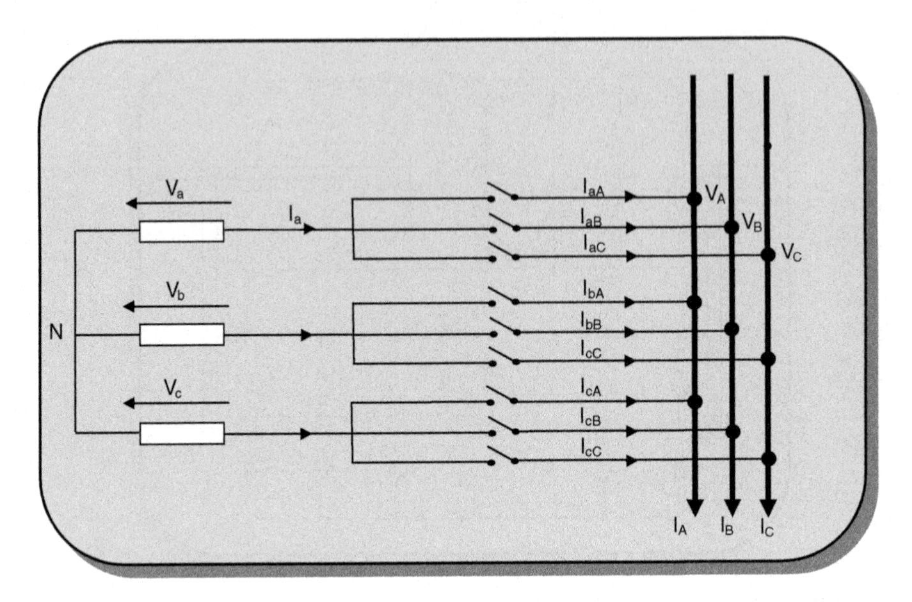

Fig. 2.19 A matrix converter circuit

the output voltage and input currents of the matrix converter can be represented by the transfer function T and the transposed T^{T} such as

$$V_o = T \cdot V_i \tag{2.48}$$

$$\begin{bmatrix} V_A \\ V_B \\ V_C \end{bmatrix} = \begin{bmatrix} S_{aA} & S_{bA} & S_{cA} \\ S_{aB} & S_{bB} & S_{cB} \\ S_{aC} & S_{bC} & S_{cC} \end{bmatrix} \cdot \begin{bmatrix} V_a \\ V_b \\ V_c \end{bmatrix} \tag{2.49}$$

$$I_i = T^{\mathrm{T}} \cdot I_o \tag{2.50}$$

The basic idea of the indirect modulation technique is to separate the control of the input currents and output voltage. The transfer function T for the matrix converter into the product of the rectifier transfer function and inverter transfer function.

$$T = R * I \tag{2.51}$$

$$\begin{bmatrix} S_{aA} & S_{aB} & S_{aC} \\ S_{bA} & S_{bB} & S_{bC} \\ S_{cA} & S_{cB} & S_{cC} \end{bmatrix} = \begin{bmatrix} S_7 & S_8 \\ S_9 & S_{10} \\ S_{11} & S_{12} \end{bmatrix} \cdot \begin{bmatrix} S_1 & S_3 & S_5 \\ S_2 & S_4 & S_6 \end{bmatrix} \tag{2.52}$$

The matrix converter model provides the basis to regard the matrix converter as a back-to-back PWM converter without any DC-link energy storage. This means the well-known space vector PWM strategies for VSI or PWM rectifier can be applied to the matrix converter.

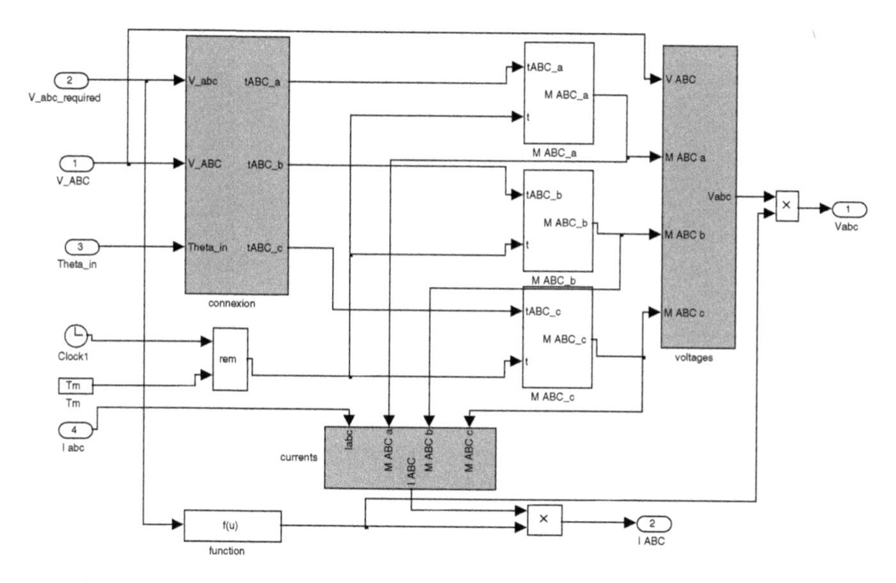

Fig. 2.20 Matrix converter model under MATLAB/Simulink

By substituting Eq. (2.52) into (2.49), we obtain

$$\begin{bmatrix} V_A \\ V_B \\ V_C \end{bmatrix} = \begin{bmatrix} S_7 & S_8 \\ S_9 & S_{10} \\ S_{11} & S_{12} \end{bmatrix} \cdot \begin{bmatrix} S_1 & S_3 & S_5 \\ S_2 & S_4 & S_6 \end{bmatrix} \cdot \begin{bmatrix} V_a \\ V_b \\ V_c \end{bmatrix} \tag{2.53}$$

$$\begin{bmatrix} V_A \\ V_B \\ V_C \end{bmatrix} = \begin{bmatrix} S_7 \cdot S_1 + S_8 \cdot S_2 & S_7 \cdot S_3 + S_8 \cdot S_4 & S_7 \cdot S_5 + S_8 \cdot S_6 \\ S_9 \cdot S_1 + S_{10} \cdot S_2 & S_9 \cdot S_3 + S_{10} \cdot S_4 & S_9 \cdot S_5 + S_{10} \cdot S_6 \\ S_{11} \cdot S_1 + S_{12} \cdot S_2 & S_{11} \cdot S_3 + S_{12} \cdot S_4 & S_{11} \cdot S_5 + S_{12} \cdot S_6 \end{bmatrix} \cdot \begin{bmatrix} V_a \\ V_b \\ V_c \end{bmatrix} \tag{2.54}$$

Therefore, the indirect modulation technique enables well-known space vector PWM to be applied for a rectifier as well as an inverter stage [4].

The model of matrix converter under MATLAB/Simulink is shown in figure (Fig. 2.20).

2.2.7 Multilevel Converter

This topology provides higher voltage and power capability and has low-switching losses, but it is a complex control with voltage unbalanced on DC-link.

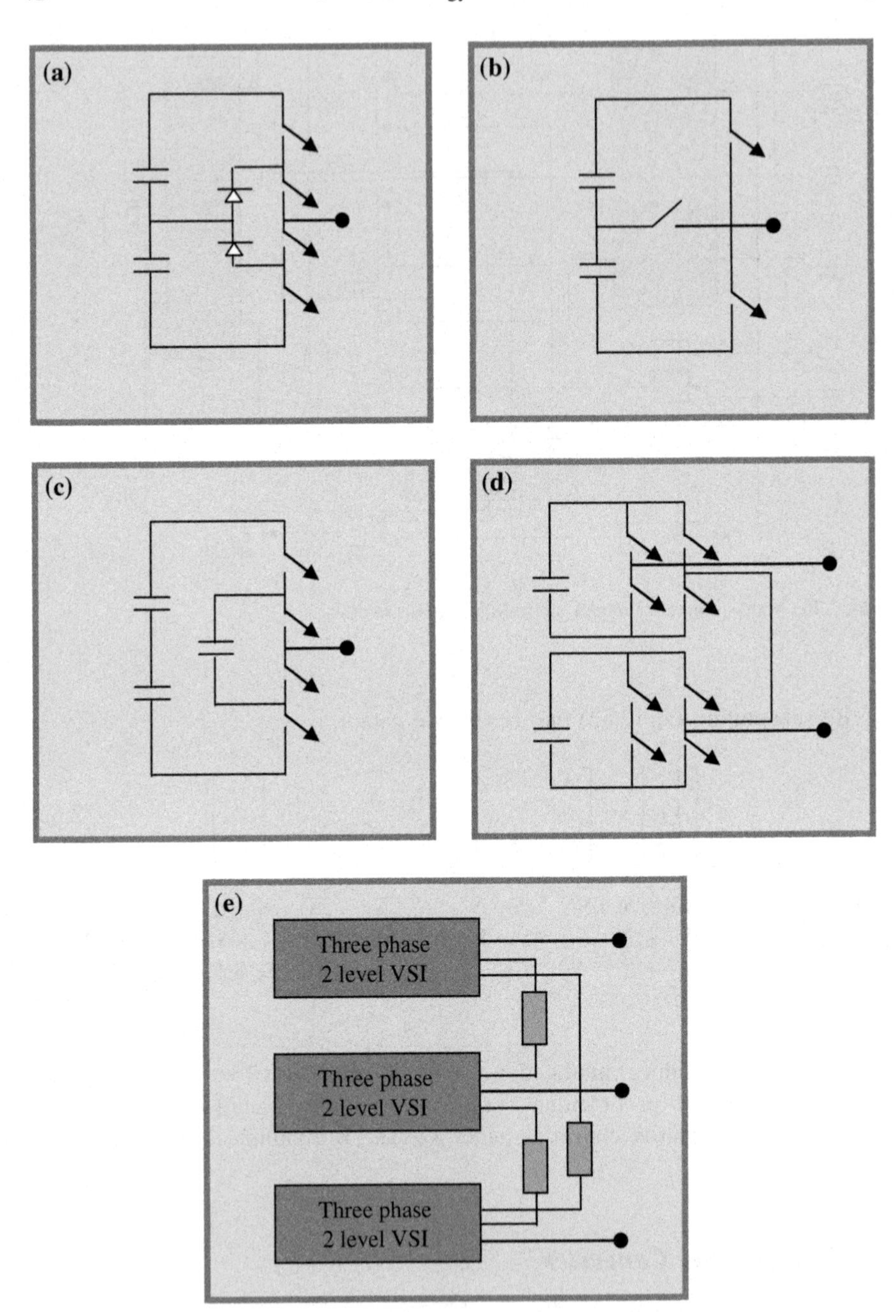

Fig. 2.21 Different multilevel topologies. **a** One inverter leg of a three level diode clamped multilevel converter. **b** One inverter leg of three level multilevel converter with bidirectional switch interconnection. **c** One inverter leg of three level flying capacitor multilevel converter. **d** One inverter leg of three level converters consisting of H bridge inverters. **e** Three level converters with three-phase inverters

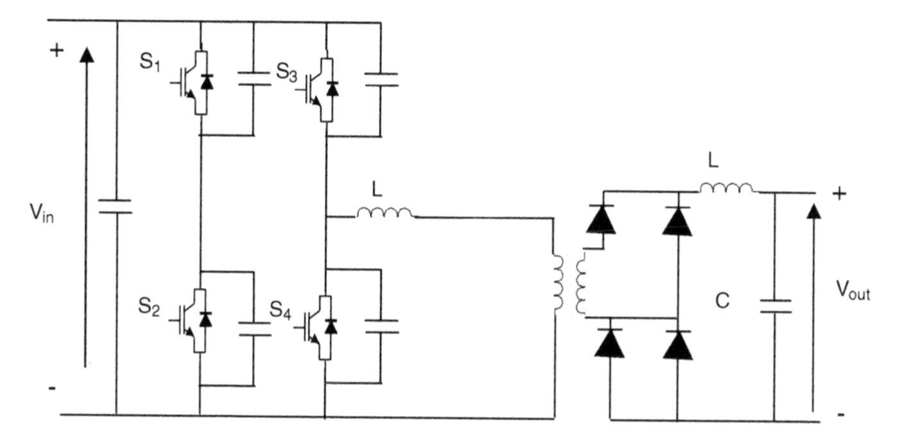

Fig. 2.22 Full-bridge converter (*FBC*)

2.2.8 DC/DC Converter

There are three topologies of DC/DC converter which be used in wind energy conversion system. They are the full bridge converter (FBC) using phase-shift control, the single active bridge (SAB) converter, and the series parallel resonant converter (SPRC) [9].

2.2.8.1 Full Bridge Converter

The most used topology in WECSs is the FBC (Fig. 2.22).

2.2.8.2 Single Active Bridge Converter

The SAB converter is shown in Fig. 2.24. It is looks similar to the FBC but due to the voltage stiff output it behaves differently and accordingly it is controlled in a different way [10] Fig. (2.23).

2.2.8.3 Resonant Converter

It is a combination of converter topologies or switching strategies that result in zero voltage and/or zero current switching. The three most used resonant converters are

- Series resonant converter SRC.
- Parallel resonant converter PRC.
- SPRC or LCC resonant converter. It is a combination of SRC and PRC topologies (Fig. 2.24).

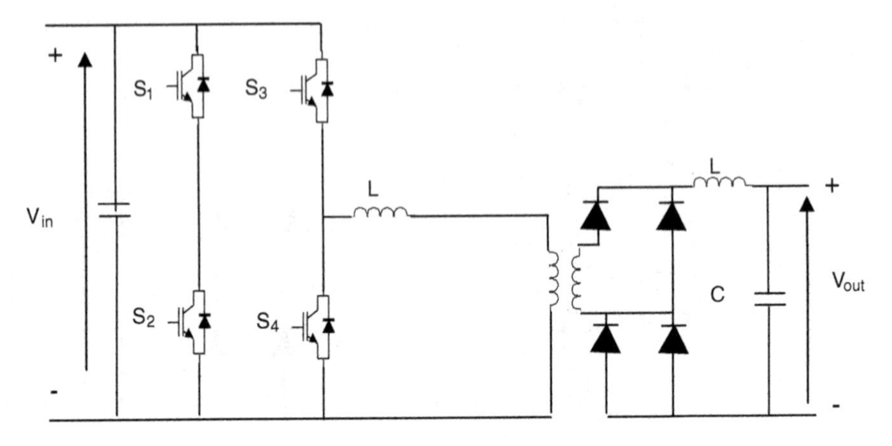

Fig. 2.23 Single active bridge (*SAB*)

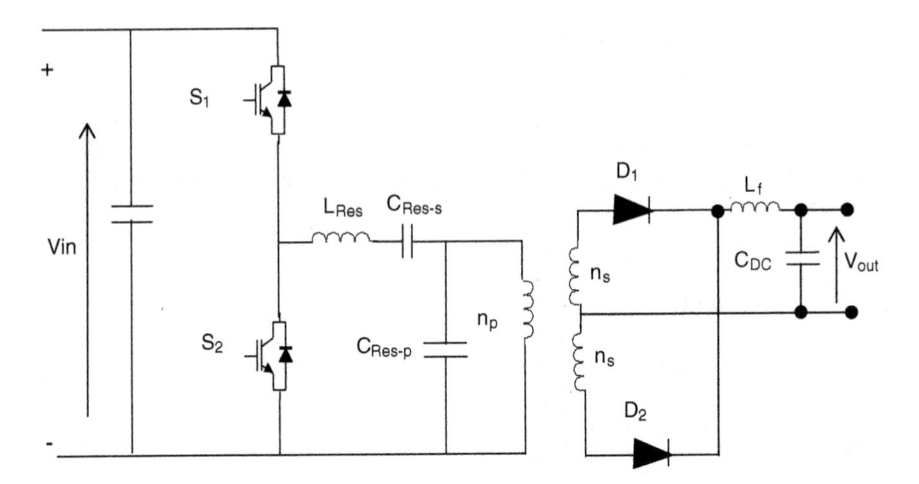

Fig. 2.24 Half-bridge series parallel resonant converter (*SPRC*)

2.2.9 Load Modeling

The most important parameters to model a load are as follows:

- the voltage level,
- kind of source (DC or AC),
- the apparent short-circuit power,
- the short-circuit angle (Fig. 2.25).

Different terms are used to characterize a grid:

- transmission grid
- distribution grid

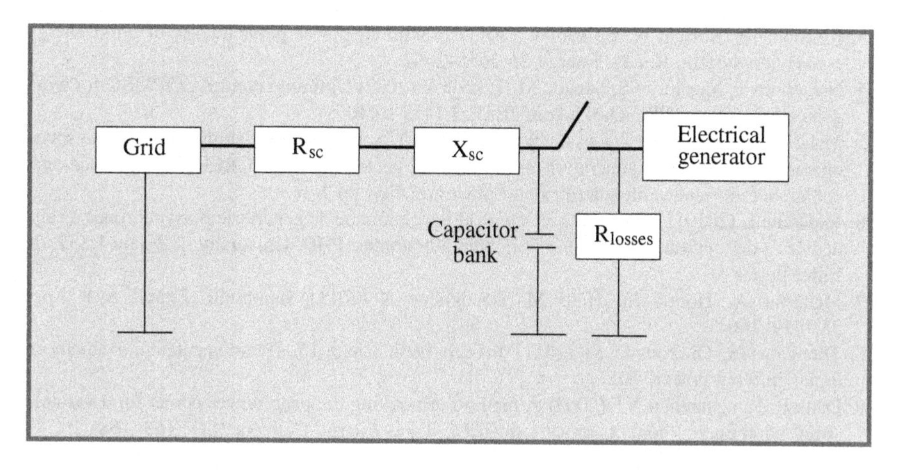

Fig. 2.25 Electrical grid model

- AC or DC grid
- stiff grid
- weak grid
- isolated grid.

2.2.10 Grid Model

We can represent electrical grid by a voltage source, short-circuit impedance, a capacitor bank, a resistance for losses, and the generator used (Fig. 2.21).

2.3 Conclusion

This chapter focuses on wind energy conversion and power electronics modeling. The energy conversion modeling has been presented with details of all its components. The different structures of converters used in wind systems are also presented. Some programs are given under MATLAB/Simulink. The global system needs to be controlled and optimized.

References

1. Abdullah MA, Yatim AHM, Tan CW, Saidur R (2012) A review of maximum power point tracking algorithms for wind energy systems. Renew Sustain Energy Rev 16:3220–3227
2. Hong YY, Lu S, Chiou CS (2009) MPPT for PM wind generator using gradient approximation. Energy Convers Manage 50:82–89

3. Kesraoui M, Korichi N, Belkadi A (2011) Maximum power point tracker of wind energy conversion system. Renew Energy 36:2655–2662
4. Nakamura T, Morimoto S, Sanada M, Takeda Y (2002) Optimum control of IPMSG for wind generation system. PPC-Osaka conf IEEE 3:1435–1440
5. Mahdi AJ, Tang WH, Jiang L, Wu QH (2010) A comparative study on variable-speed operations of a wind generation system using vector control. ICREPQ'10: international conference on renewable energies and power quality, pp 1–6
6. Idjdarene K (2010) Contribution à l'étude et la commande de génératrices asynchrones à cage dédiées à des centrales électriques éoliennes autonomes. PHD, University of Bejaia-USTL de Lille, Bejaia
7. Meharrar A, Tioursi M, Hatti M, Boudghène A (2011) Stambouli. Expert Syst Appl 38:7659–7664
8. Thongam JS, Ouhrouche M (2011) InTech book Chap. 15, (Fundamental and advanced topics in wind power, 2011)
9. Datta R, Ranganathan VT (2003) A method of tracking the peak power points for a variable speed wind energy conversion system. IEEE Trans Energy Convers 18(1):163–168
10. Munteanu I, Bratu AI, Cutululis NA, Ceanga E (2008) Optimal control of wind energy systems. Springer-Verlag, London

Chapter 3
Optimisation of Wind System Conversion

3.1 Introduction to Optimization Algorithms

The output power of wind energy system varies depending on the wind speed. Due to the nonlinear characteristic of the wind turbine, it is difficult to maintain the maximum power output of the wind turbine for all wind speed conditions. Therefore, over the years, several maximum power point tracking (MPPT) algorithms have been developed to track the maximum power point of the wind turbine [1–34].

3.2 Maximum Power Point Tracking Algorithms

3.2.1 Perturb and Observe (P&O) Technique or Hill Climb Searching (HCS)

Hill climb searching (HCS) or perturbation and observation (P&O) control is usually used for the peak power of the wind turbine that will maximize the extracted energy. This control efforts to climb the $P_m(\omega_m)$ curve in the direction of increasing P_m, by varying the rotational speed periodically with a small incremental step in order to reduce the oscillation around the MPP. The P&O algorithm compares the power previously delivered after one disturbance. A flowchart of this method is shown in Fig. 3.1.

The principle can be described as follows: As shown in Fig. 3.2, in the ascending phase of the $P_m(\omega_m)$ characteristics and considering a positive change of the rotational speed, the tracker generates a positive change $\Delta\omega_m > 0$, which results in an increase in the delivered mechanical power and change of the operating point Xi $(i = 1, 2, \ldots, n - 1)$. In this case, the rotational speed and the P_m power increase up to a new point Xi + 1. Similar steps with opposite direction can be done in the case of a decrease in the mechanical power, by setting $X = \omega_m$; the instantaneous rotational speed of the wind turbine follows the maximum power point according to a predetermined rotational speed and power values. Under these

D. Rekioua, *Wind Power Electric Systems*, Green Energy and Technology,
DOI: 10.1007/978-1-4471-6425-8_3, © Springer-Verlag London 2014

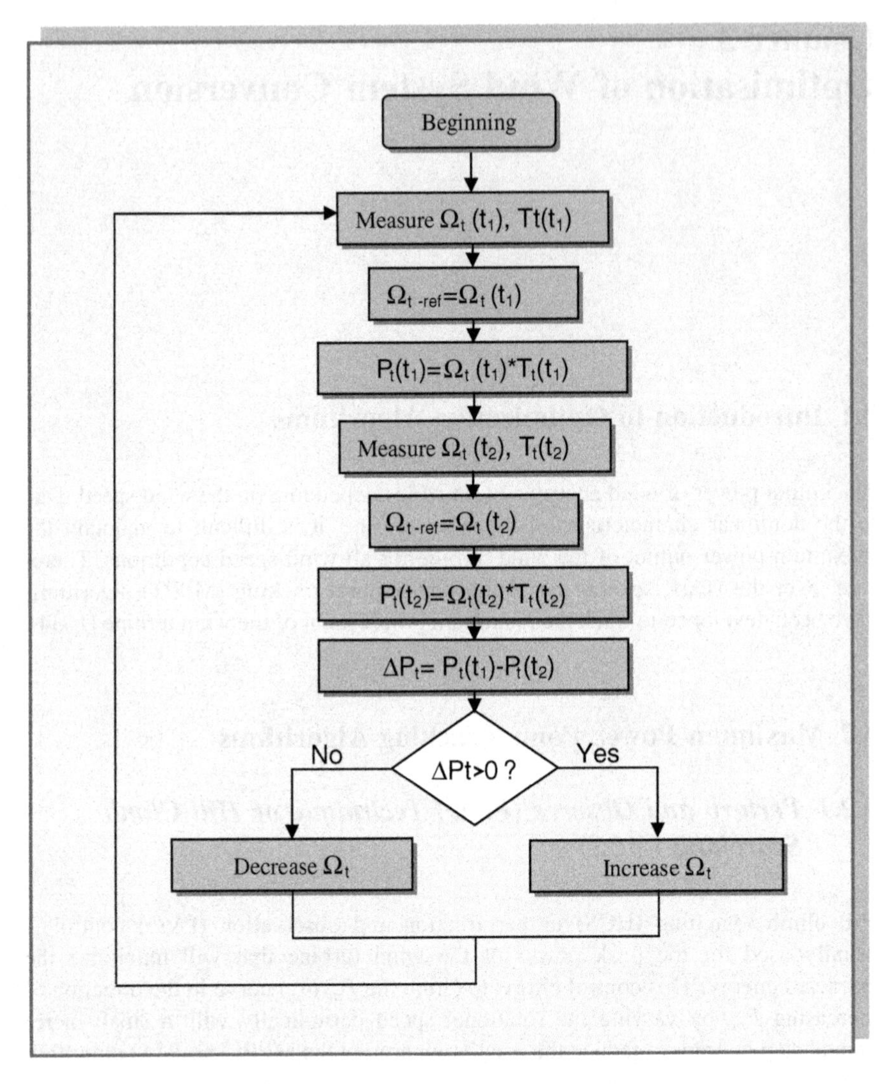

Fig. 3.1 Flowchart of the P&O method

conditions, the tracker seeks the MPP permanently. At specified wind speed, the desired mechanical power is the solution of the nonlinear equation given by $(dP_m/d\omega_m = 0)$.

The controlling rule for adjusting the step size varies from one group of studies to another, depending on the disturbed variable [9, 15, and 30]. The magnitude of the step size is the main factor determining the amplitude of oscillations that allows the convergence rate to the final response. Nevertheless, a larger disturbance will lead to a higher value of oscillation amplitude around the peak point. In this algorithm, there is a trade-off between the rate of response and the amount

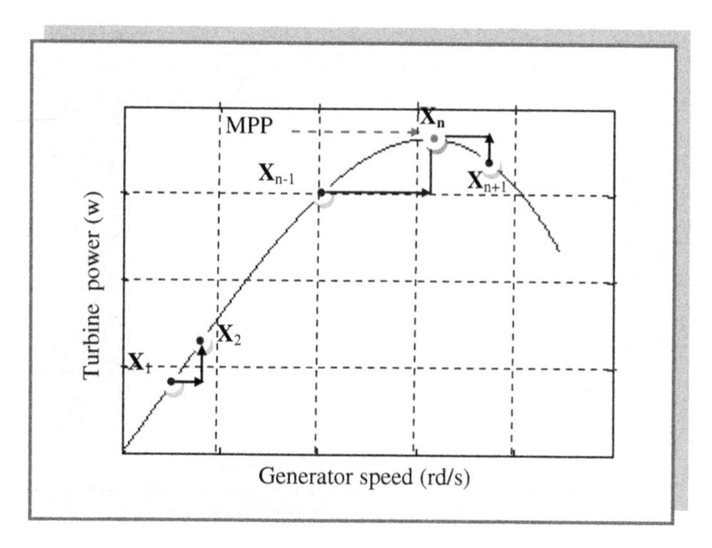

Fig. 3.2 The principle of the P&O MPPT

of oscillations under steady state conditions. To overcome this trade-off, the step size of varying amplitude can be applied. The step-size amplitude can be determined according to power variations based on the previously applied disturbance. Therefore, larger step-size amplitude is selected when power is far from MPP due to the larger magnitude of $P_m(\omega_m)$ slope and small amplitude is selected when power is close to MPP. The step size is continually decreased until it approaches zero in the aim to drive the operating point to settle down at the MPP.

The speed step is computed as:

$$\Delta\omega_m(k) = C.\frac{dP_m(k)}{d\omega_m(k)}$$

And the reference generator speed is computed as:

$$\omega_m^*(k+1) = \omega_m(k) + \Delta\omega_m(k) \tag{3.1}$$

where: $\omega_m(K)$ and $\omega_m(K+1)$ are generator speed values at sampling time (k) and ($k+1$), respectively, and (C) is the step change (Fig. 3.3).

3.2.1.1 Application Under Matlab/Simulink

We make an application of the P&O method in a wind system. The block diagram under Matlab/Simulink can be represented in Fig. 3.4.

The turbine model is obtained using equations given in Chap. 1. We can implement it under Matlab/Simulink in Fig. 3.5.

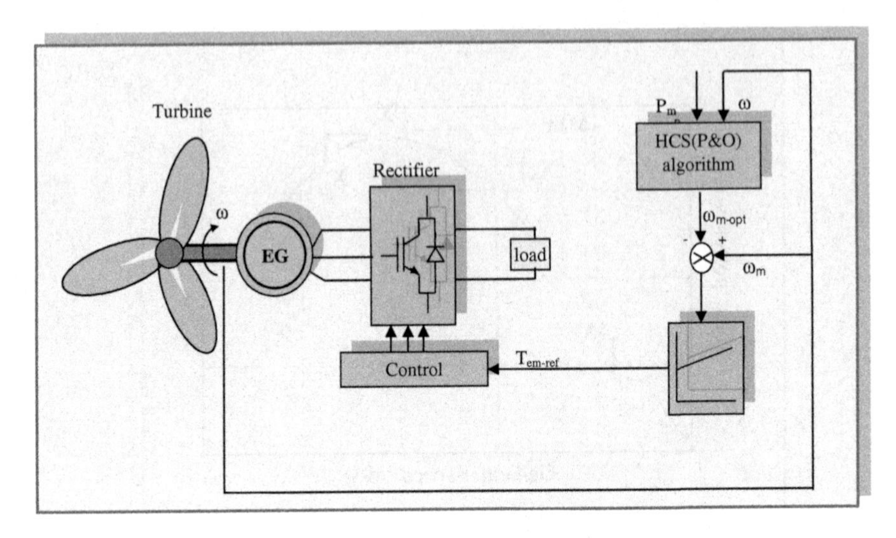

Fig. 3.3 Hill climb search control of wind energy conversion system

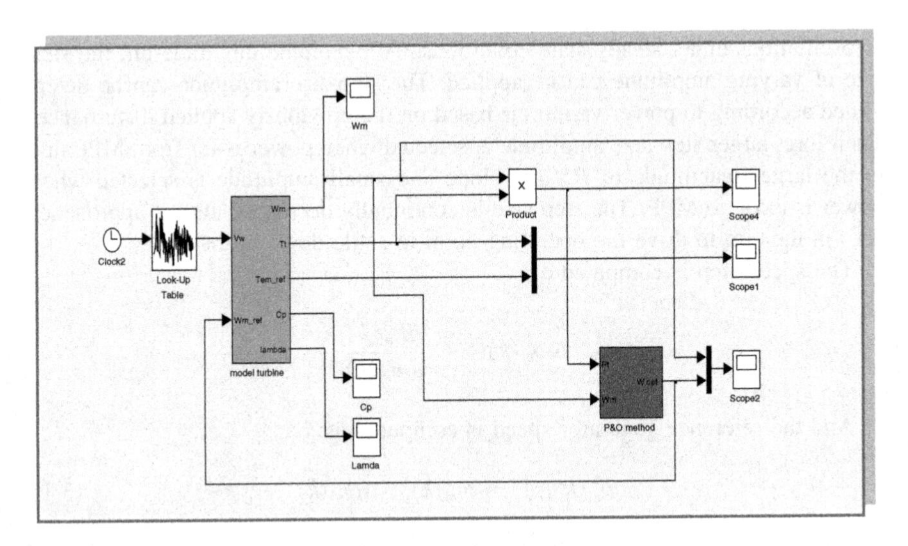

Fig. 3.4 Block diagram of WTCS with P&O method

The P&O method is represented in Fig. 3.6.

We make an application with a chosen wind speed profile (Fig. 3.7). We note that the power coefficient C_p kept constant (Fig. 3.8) whatever the wind variations and the reference torque follows the turbine torque (Figs. 3.9 and 3.10).

We note also that the turbine speed follows the wind profile (Fig. 3.11).

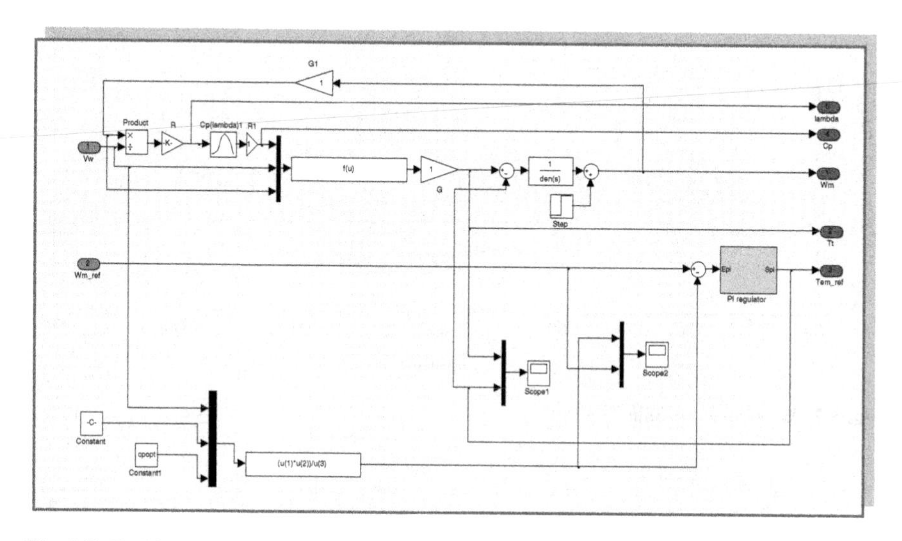

Fig. 3.5 Turbine model under Matlab/Simulink

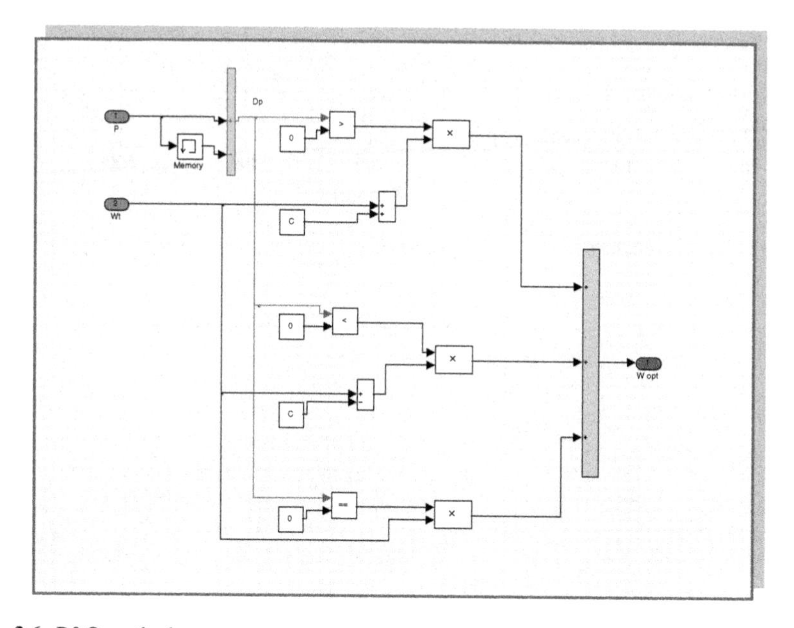

Fig. 3.6 P&O method

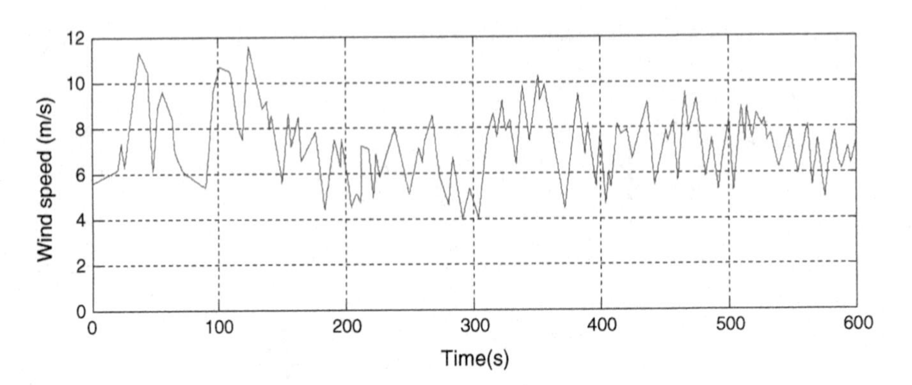

Fig. 3.7 Wind speed profile

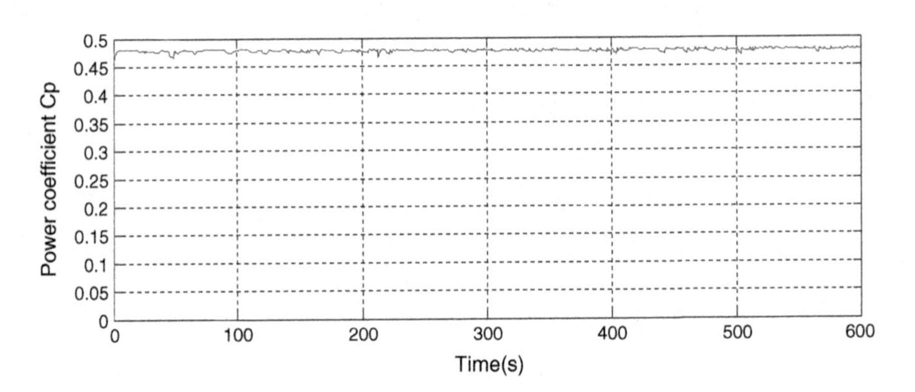

Fig. 3.8 Power coefficient

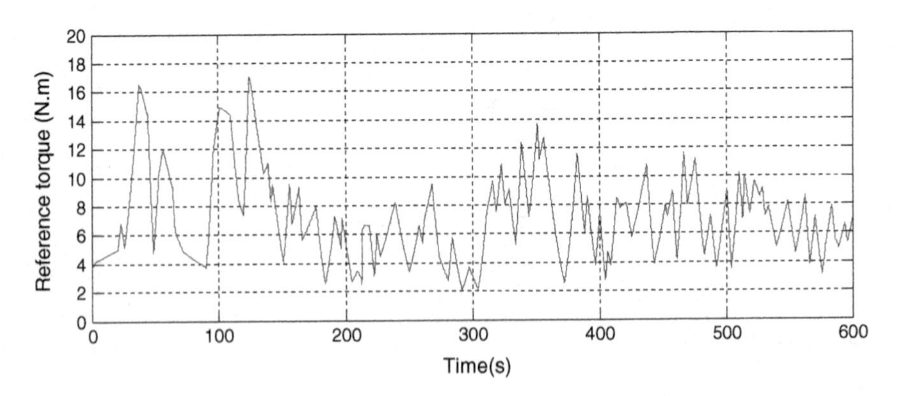

Fig. 3.9 Reference torque

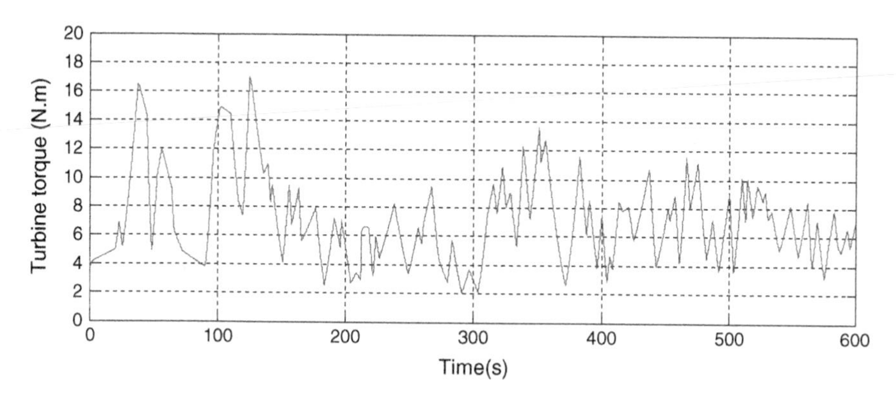

Fig. 3.10 Turbine torque

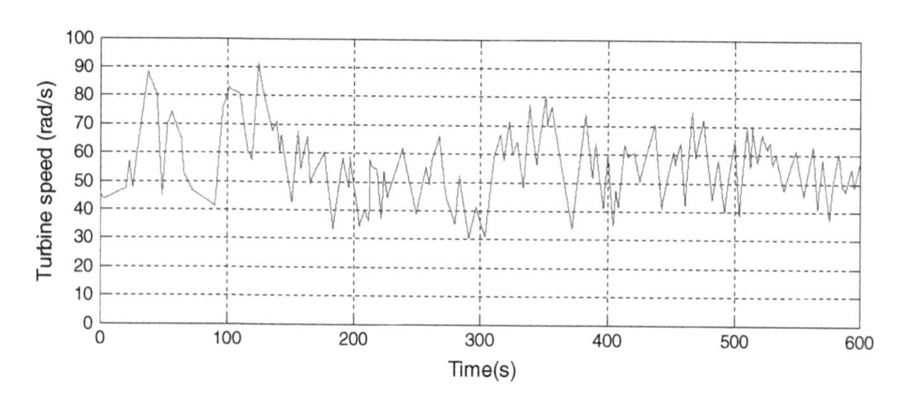

Fig. 3.11 Turbine speed

3.2.2 Tip Speed Ratio Method

In order to have maximum possible power, the turbine should always operate at λ_{opt}. The tip speed ratio (TSR) control method regulates the tip speed ratio to maintain it to an optimal value, at which the rotational speed is optimum and the power extracted is maximum. This control requires the knowledge of wind speed, the turbine speed, and the reference optimal point of the TSR which can be determined experimentally or theoretically [3]. The comparison of the TSR reference with the actual value feeds this difference to the controller and gives the reference power. Figure 3.12 shows the block diagram of a WECS with TSR method.

We make an application of the TSR method in a wind system. The block diagram under Matlab/Simulink can be represented in Fig. 3.13.

The TSR method is represented in Fig. 3.14.

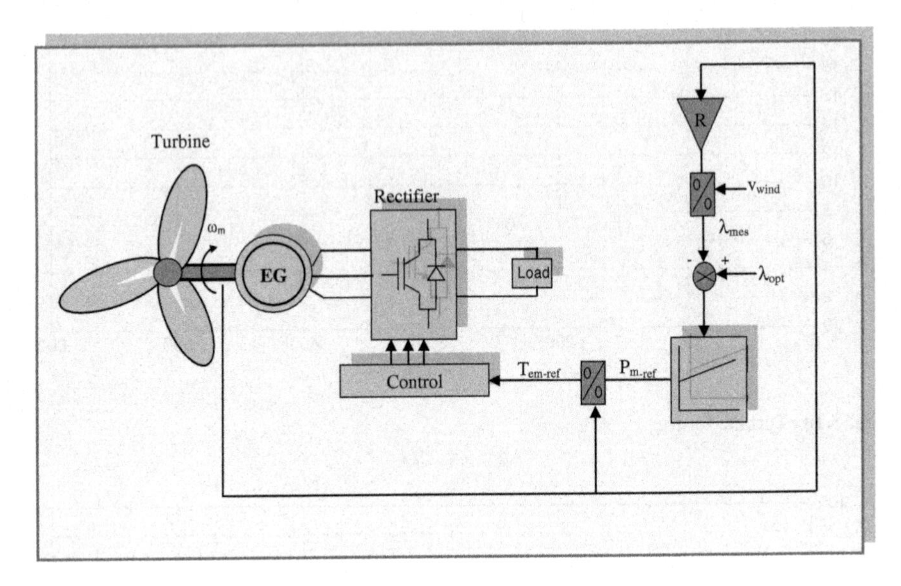

Fig. 3.12 Tip speed ratio method of wind energy conversion system

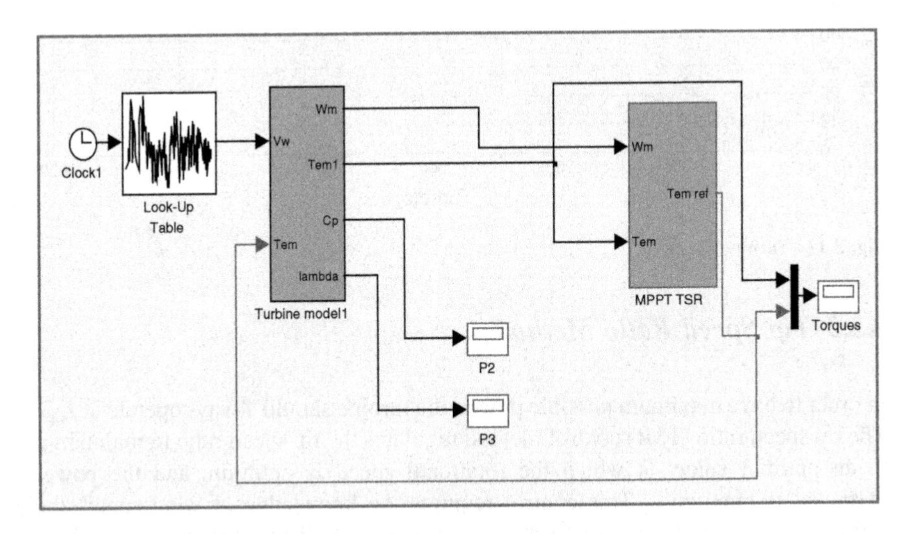

Fig. 3.13 Block diagram of WTCS with TSR MPPT method

3.2.3 Power Signal Feedback (PSF) Method

Power signal feedback method generates a reference power signal to maximize the output power. However, it requires the knowledge of the wind turbine and maximum power curve which can be obtained from the experimental results or simulations [21]. Then, the data points for maximum turbine power and the

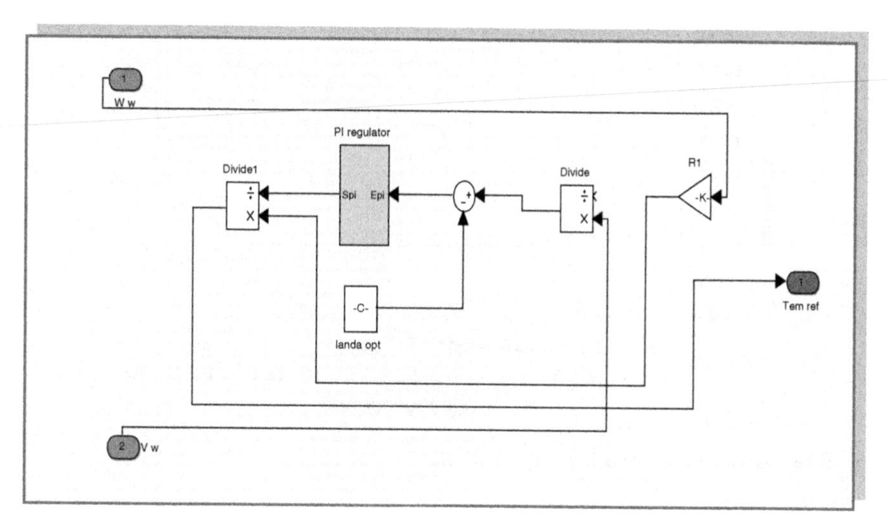

Fig. 3.14 TSR method

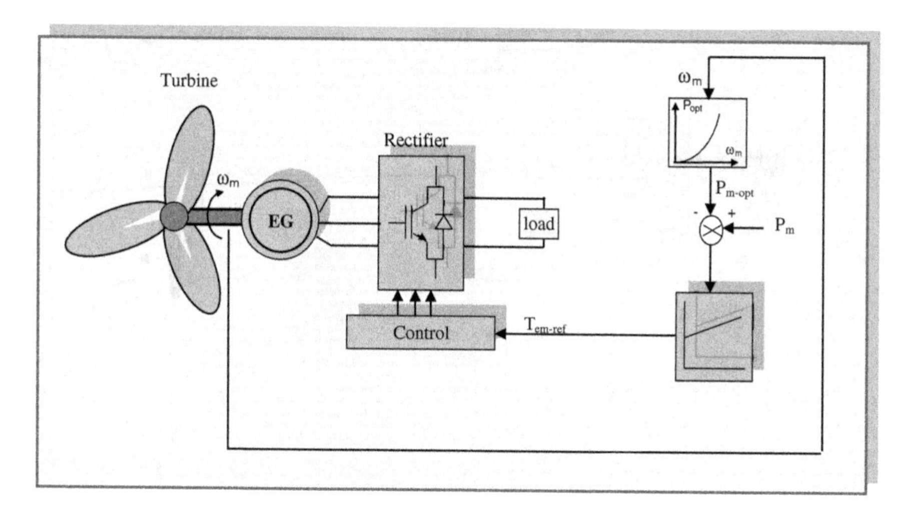

Fig. 3.15 Power signal feedback control of wind energy conversion system

corresponding wind turbine speed must be recorded in a lookup table [8, 11]. The PSF control method regulates the turbine power to maintain it to an optimal value, so that the power coefficient C_p is always at its maximum value corresponding to the optimum tip speed ratio. Figure 3.15 shows the block diagram of a WECS with TSR control.

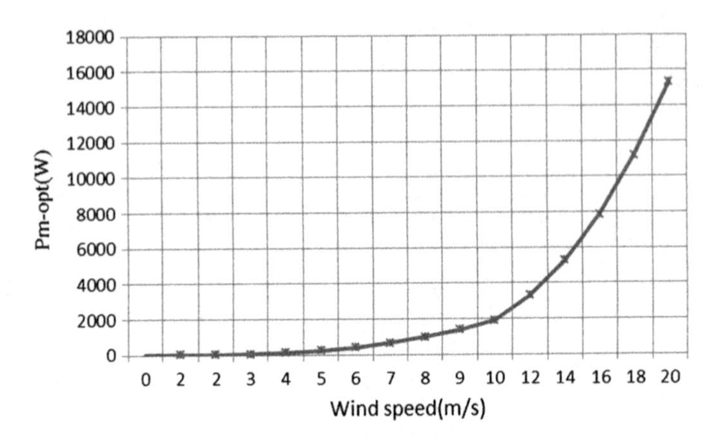

Fig. 3.16 Maximum power using polynomial curve

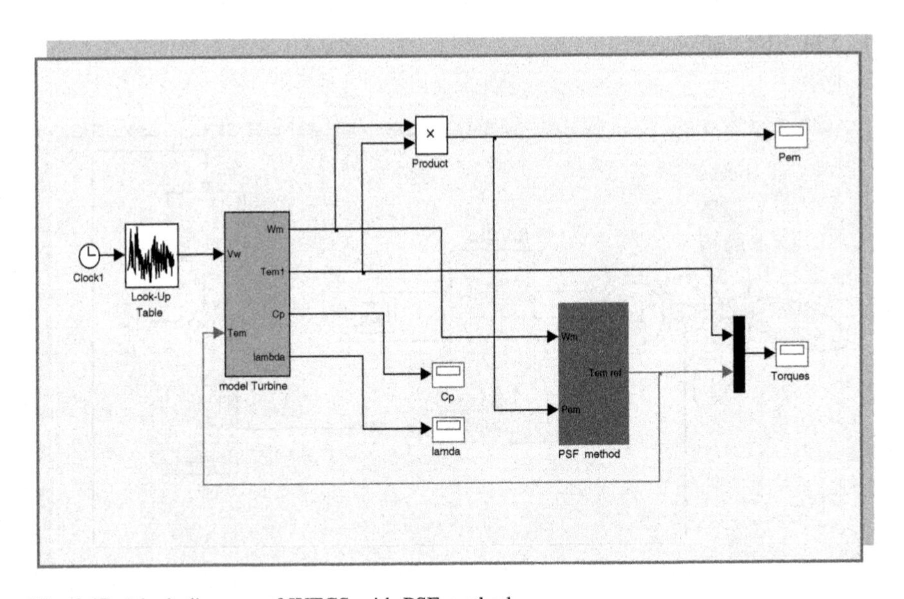

Fig. 3.17 Block diagram of WTCS with PSF method

Maximum power output at different wind speeds can be expressed using polynomial curve fit

$$P_{m-\text{opt}} = -0.3 + 1.08 \cdot V_{Wv}^{3} - 0.125 \cdot V_W^2 + 0.842 \cdot V_W^3 \tag{3.2}$$

We can represent it in Fig. 3.16.

The block diagram of WTCS with PSF method under Matlab/Simulink is represented in Fig. 3.17.

The PSF method is represented in Fig. 3.18.

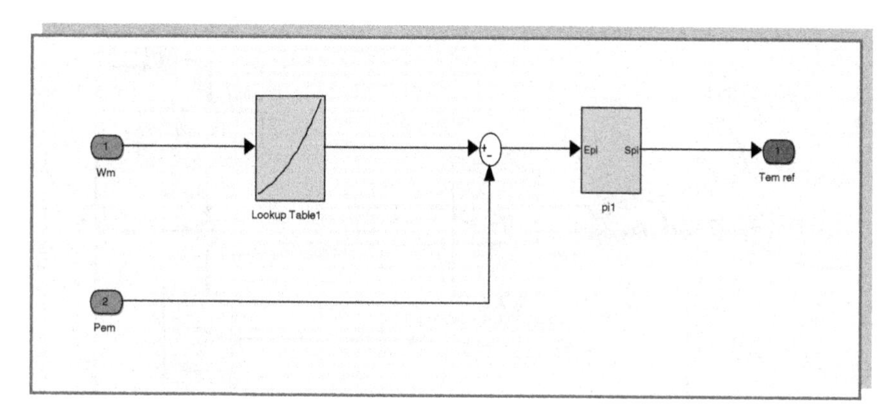

Fig. 3.18 PSF method

3.3 Optimal Torque Control

Optimal torque control (OTC) is a slight variant of PSF control [14]. It adjusts the generator torque to its optimal at different wind speeds. However, it requires the knowledge of turbine characteristics (C_{pmax} and λ_{opt}).

We have:

$$v_{\text{wind}} = \frac{R \cdot \omega_t}{\lambda} \tag{3.3}$$

We obtain the power function of rotational turbine speed

$$P_t(\omega_t) = \frac{1}{2} \cdot \frac{C_p(\omega_t) \cdot \rho \cdot \pi \cdot R^4}{\lambda^3(\omega_t)} \cdot \omega_t^3 \tag{3.4}$$

With:

$$P_t(\omega_t) = T_{\text{em}} \cdot \omega_t \tag{3.5}$$

$$C_t \cdot \omega_t = \frac{1}{2} \cdot C_{\text{p}}(\lambda) \cdot \rho \cdot \pi \cdot R^2 \cdot v_{\text{wind}}^3 \tag{3.6}$$

Then:

$$T_{\text{em}} = \frac{1}{2} \cdot \frac{C_p(\omega_t) \cdot \rho \cdot \pi \cdot R^4}{\lambda^3(\omega_t)} \cdot \omega_t^2 \tag{3.7}$$

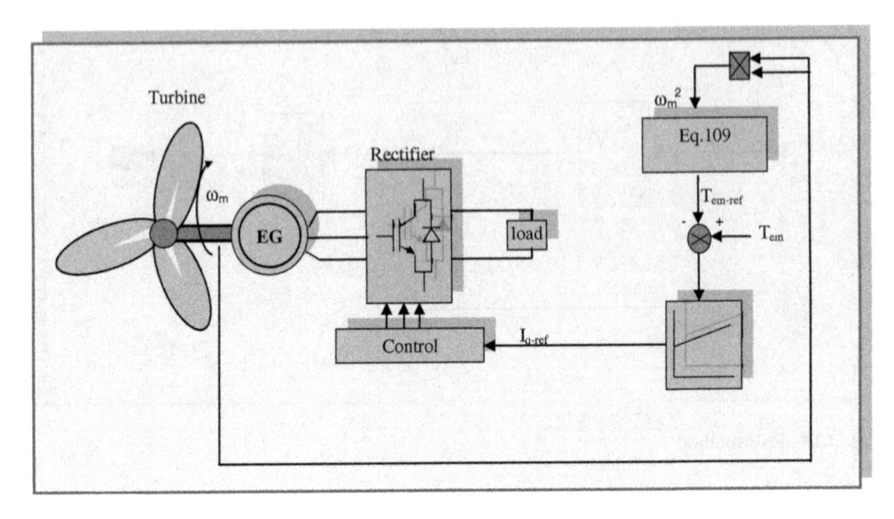

Fig. 3.19 Optimal torque control of wind energy conversion system

Assuming optimal conditions, power value, speed, and optimal torque are given by the following relations:

$$P_{\text{opt}} = \frac{1}{2} \cdot C_{p-opt} \cdot \rho \cdot \pi \cdot R^2 \cdot v_{\text{wind}}^3 \tag{3.8}$$

$$\omega_{\text{opt}} = \frac{v_{\text{wind}} \cdot \lambda_{\text{opt}}}{R} \tag{3.9}$$

$$T_{\text{em-opt}} = T_{\text{em-ref}} = K_{\text{opt}} \cdot \omega_{\text{opt}}^2 \tag{3.10}$$

With:

$$K_{\text{opt}} = \frac{1}{2} \cdot \frac{C_{\text{p-opt}}(\omega_t) \cdot \rho \cdot \pi \cdot R^4}{\lambda_{\text{opt}}^3(\omega_t)} \tag{3.11}$$

Figure 3.19 shows the block diagram of a WECS with OTC.

Application sous Matlab/Simulink:
We make an application of the OTC method in a wind system. The block diagram under Matlab/Simulink can be represented in Fig. 3.20.

The OTC method is represented in Fig. 3.21.

We make an application with a chosen wind speed profile (Fig. 3.22). We remark that the power coefficient C_p kept constant (Fig. 3.23) whatever the wind variations and the reference torque follows the turbine torque (Figs. 3.24 and 3.25).

A comparison between the three most used MPPT methods (P&O, PSF, and TSR), has been made to show their performances (Figs. 3.26, 3.27 and 3.28).

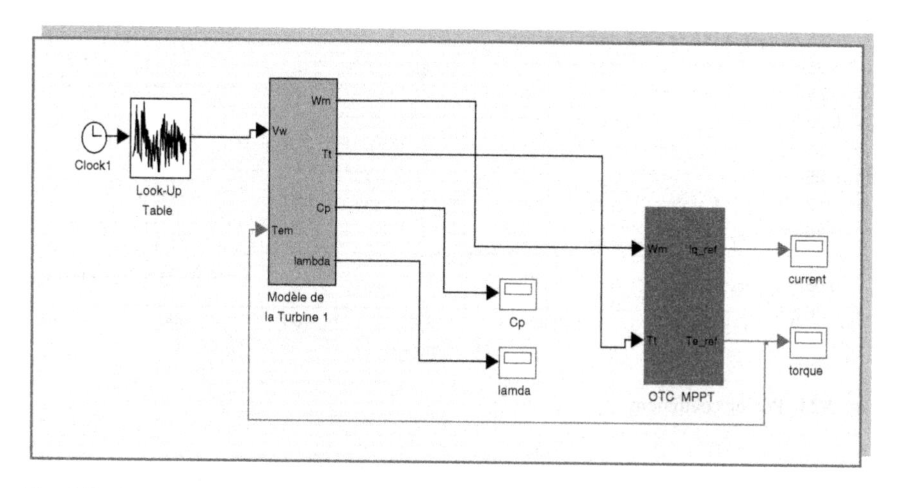

Fig. 3.20 Block diagram of WTCS with OTC method

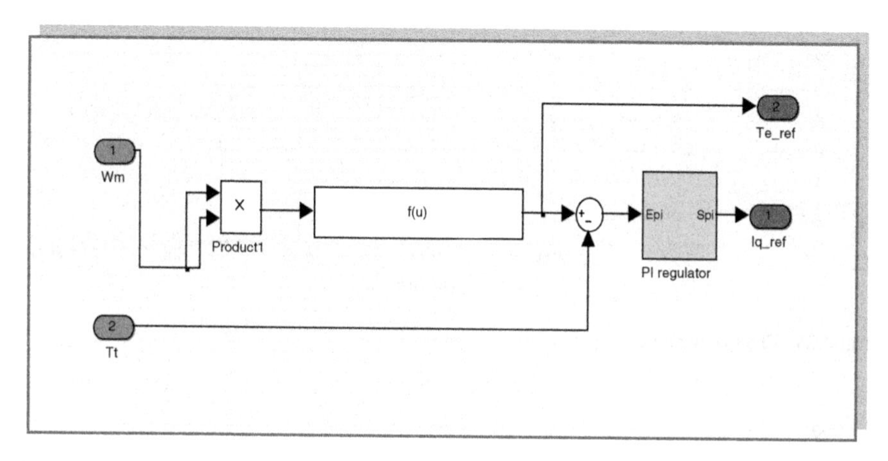

Fig. 3.21 OTC method

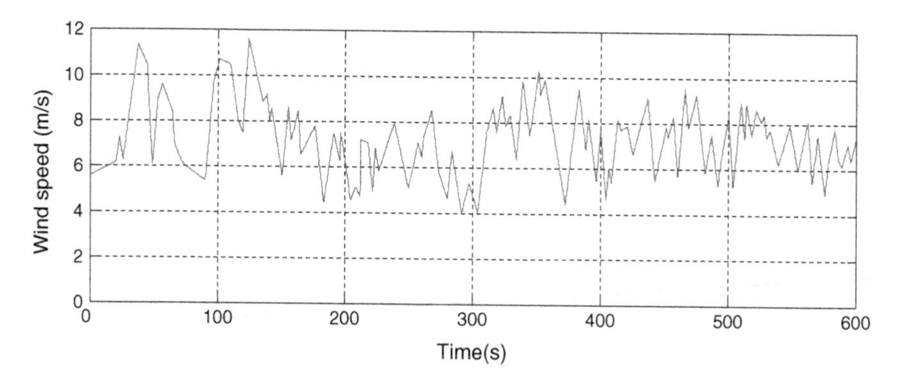

Fig. 3.22 Wind speed profile

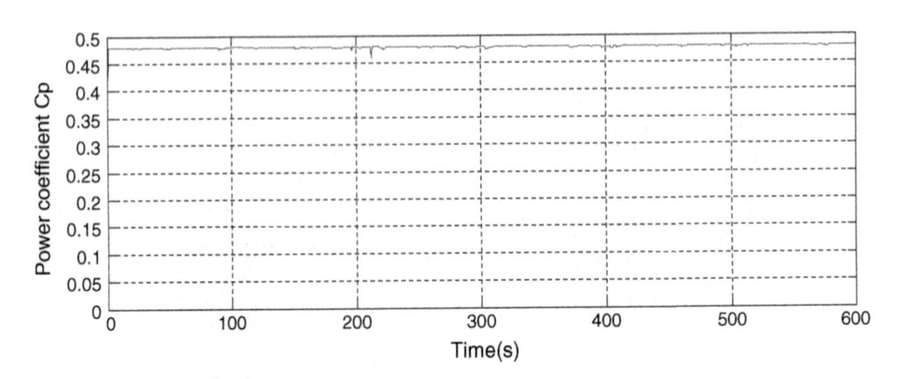

Fig. 3.23 Power coefficient

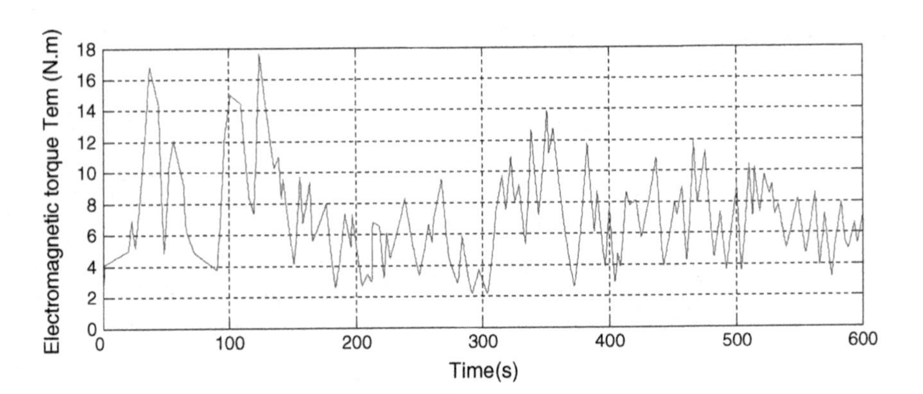

Fig. 3.24 Electromagnetic torque turbine response

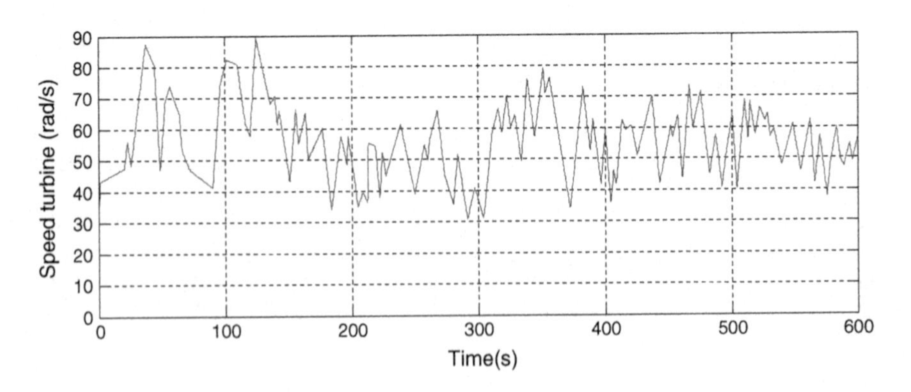

Fig. 3.25 Speed turbine response

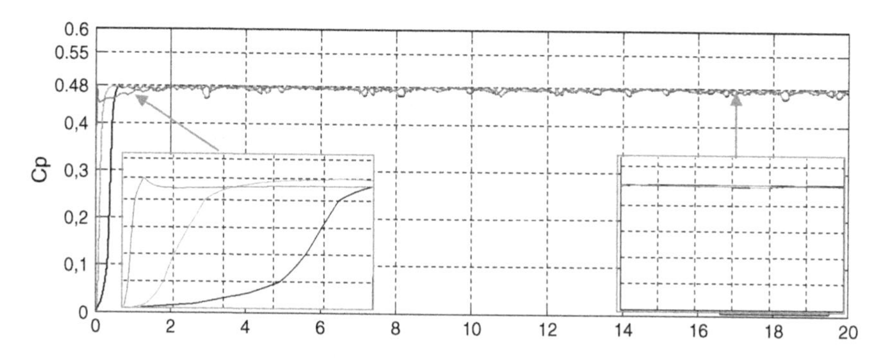

Fig. 3.26 Power coefficient of three methods (P&O, PSF, and TSR)

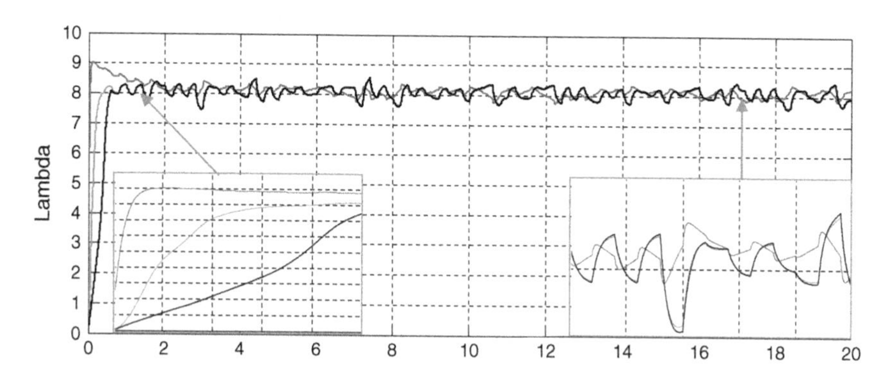

Fig. 3.27 TSR of three methods (P&O, PSF, and TSR)

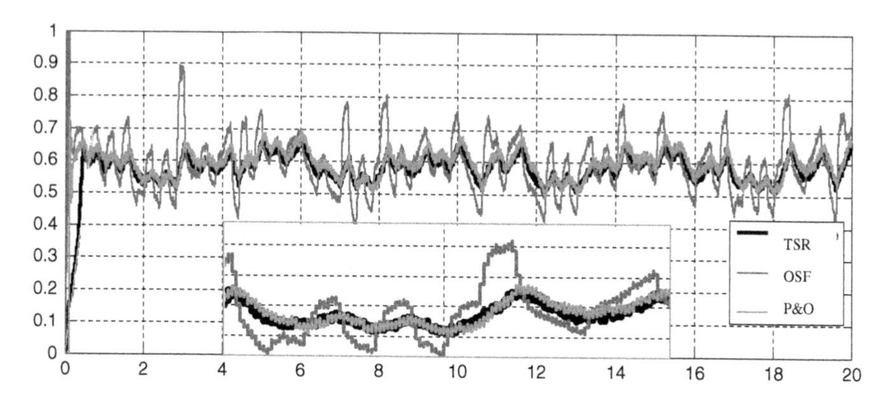

Fig. 3.28 Efficiency of three methods (P&O, PSF, and TSR)

3.4 Sliding Mode Control

The advantages of sliding mode control are various and important: high precision, good stability, simplicity, invariance, robustness, etc. This allows it to be particularly suitable for systems with imprecise model. Often it is better to specify the dynamics system during the convergence mode. In this case, the controller structure has two parts, one represents the dynamics during the sliding mode and the other represents the discontinuous dynamics system during the convergence mode.

The design of the control can be obtained in three important steps, and each step is dependent on another one:

- **The choice of surface**

For a system defined by the following equation, the vector of surface has the same dimension as the control vector (u).

$$\dot{x} = A(x,t) \cdot x + B(x,t) \cdot u \tag{3.12}$$

We find in scientific researches different forms of the sliding surface, and each surface has better performance for a given application. In general, we choose a nonlinear surface. The nonlinear form is a function of the error on the controlled variable (x), and it is given by:

$$S(x) = \left(\frac{\partial}{\partial t} + \lambda_x\right)^{r-1} \cdot e(x) \tag{3.13}$$

where $e(x) = \hat{x} - x$ is the difference between the controlled variable x and its reference $\hat{x}$, λ_x is a positive constant, r is the number of times to derive the surface to obtain the control, and x is the controlled variable.

The purpose of the control is to maintain the surface to zero. This one is a linear differential equation with a unique solution $e(x) = 0$ for a suitable choice of parameter λ_x with respect to the convergence condition.

- **The establishment of the invariance conditions**

The conditions of invariance and convergence criteria have different dynamics that allow the system to converge to the sliding surface and stay there regardless of the disturbance: There are two considerations to ensure the convergence mode.

- **The discrete function switching**

This is the first convergence condition. We have to give to the surface a dynamic converging to zero. It is given by:

$$\begin{cases} \dot{S}(x) > 0 & \text{if } S(x) < 0 \\ \dot{S}(x) < 0 & \text{if } S(x) > 0 \end{cases} \tag{3.14}$$

It can be written as:

$$\dot{S}(x) \cdot S(x) < 0 \tag{3.15}$$

- **Lyapunov function**

The Lyapunov function is a positive scalar function for the state variables of the system. The idea is to choose a scalar function to ensure the attraction of the variable to be controlled to its reference value.

We define the Lyapunov function as follows:

$$V(x) = \frac{1}{2} \cdot S^2(x) \tag{3.16}$$

The derivative of this function is:

$$\dot{V}(x) = S(\mathrm{x}) \cdot \dot{S}(\mathrm{x}) \tag{3.17}$$

The function will decrease, if its derivative is negative. This is checked only if the condition (Eq. 3.25) is verified.

- **Determination of the control law**

The structure of a sliding mode controller consists of two parts. The first one concerns the exact linearization (u_{eq}), and the second one concerns the stabilization (u_n).

$$u = u_{\text{eq}} + u_n \tag{3.18}$$

where u_{eq} corresponds to the control. It serves to maintain the variable control on the sliding surface.

u_n is the discrete control determined to check the convergence condition (Eq. 3.15).

We consider a system defined in state space, and we have to find analogical expression of the control (u).

$$\dot{S}(x) = \frac{\partial S}{\partial t} = \frac{\partial S}{\partial x} \cdot \frac{\partial x}{\partial t} \tag{3.19}$$

Substituting Eqs. 3.17 and 3.13 in Eq. 3.19, we obtain:

$$\dot{S}(x) = \frac{\partial S}{\partial x} \cdot \left(A(x,t) + B(x,t) \cdot u_{eq}\right) + \frac{\partial S}{\partial x} \cdot B(x,t) \cdot u_n \tag{3.20}$$

We deduce the expression of the equivalent control:

$$u_{\text{eq}} = -\left(\frac{\partial S}{\partial t} \cdot B(x,t)\right)^{-1} \cdot \frac{\partial S}{\partial t} \cdot A(x,t) \tag{3.21}$$

For the equivalent control can take a finite value, it must:

$$\frac{\partial S}{\partial x} \cdot B(x,t) \neq 0 \tag{3.22}$$

In the convergence mode and replacing the equivalent control by its expression, we find the new expression of the surface derivative:

$$\dot{S}(x,t) = \frac{\partial S}{\partial x} \cdot B(x,t) \cdot u_n \tag{3.23}$$

And the condition expressed by Eq. 3.14 becomes:

$$S(x,t) \cdot \frac{\partial S}{\partial x} \cdot B(x,t) \cdot u_n < 0 \tag{3.24}$$

The simplest form that can take the discrete control is as follows:

$$u_n = k_s \cdot \text{sign}(S(x,t)) \tag{3.25}$$

where the sign of k_s must be different from that of $\frac{\partial S}{\partial x} \cdot B(x,t)$.

3.5 Fuzzy Logic Controller Technique

Fuzzy logic controller (FLC) is introduced to determine the operating point corresponding to the maximum power for different wins speed. In this case, inputs of the FLC are power variation (ΔP_{wind}) and speed variation ($\Delta\Omega_t$). The output is reference voltage variation ($\Delta\Omega_{t,\text{ref}}$). In order to converge toward the optimal point, rules are relatively simple to establish. These rules depend on the variations of power ΔP_{wind} and voltage $\Delta\Omega_t$. In accordance with Table 3.1, if the power (P_{wind}) increased, the operating point should be increased as well. However, if the power (P_{wind}) decreased, the speed ($\Omega_{t,\text{ref}}$) should do the same.

Table 3.1 Fuzzy rule table

ΔP_{wind} / $\Delta\Omega_t$	BN	MN	SN	Z	SP	MP	BP
BN	BP	BP	MP	Z	MN	BN	BN
MN	BP	MP	SP	Z	SN	MN	BN
SN	MP	SP	SP	Z	SN	SN	MN
Z	BN	MN	SN	Z	SP	MP	BP
SP	MN	SN	SN	Z	SP	SP	MP
MP	BN	MN	SN	Z	SP	MP	BP
BP	BN	BN	MN	Z	MP	BP	BP

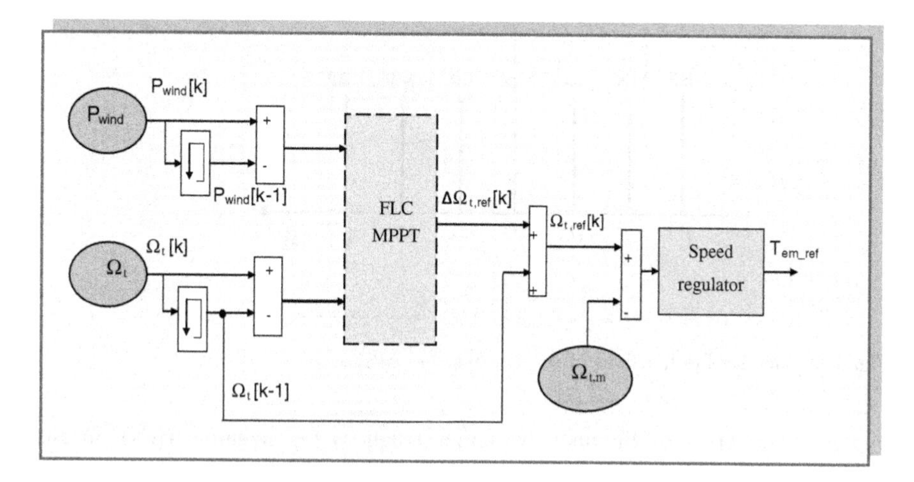

Fig. 3.29 Structure of fuzzy controller MPPT wind applied to the system

From these linguistic rules, the MPPT algorithm contains measurement of variation of wind power ΔP_{wind} and variation of turbine speed $\Delta\Omega_t$ proposes a variation of $\Delta\Omega_{t,\text{ref}}$ according to Eq. 3.26.

$$\begin{cases} \Delta P_{\text{wind}} = P_{\text{wind}}(k) - P_{\text{wind}}(k-1) \\ \Delta\Omega_t = \Omega_t(k) - \Omega_t(k-1) \\ \Omega_{t-\text{ref}}(k) = \Omega_t(k-1) + \Omega_{t-\text{ref}}(k) \end{cases} \tag{3.26}$$

where $P_{\text{wind}}(k)$ and $\Omega_t(k)$ are the power and speed of the turbine at sampled times (k), and $\Delta\Omega_{t,\text{ref}}$ (k) is the instant of reference speed.

The block FLC includes three functional blocks: fuzzification, fuzzy rule algorithm, and defuzzification (Fig. 3.29).

Figure 3.30 shows the membership function of input and output variables in which membership functions of input variables ΔP_{wind} and $\Delta\Omega_t$ are triangular and have seven fuzzy subsets. Seven fuzzy subsets are considered for membership functions of the output variable $\Delta\Omega_{t,\text{ref}}$.. These inputs and output variables are

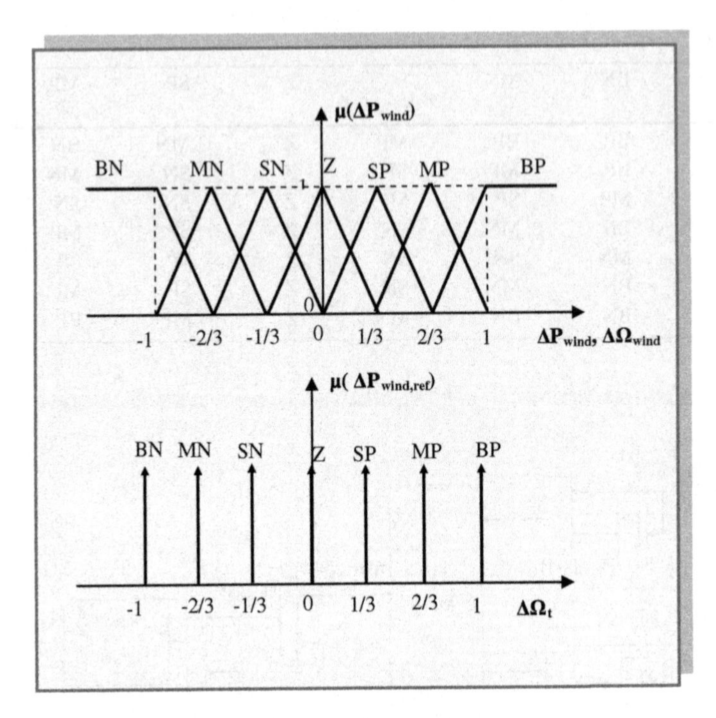

Fig. 3.30 Membership functions of ΔP_{wind}, $\Delta\Omega_t$, and $\Delta P_{\textbf{wind},\text{ref}}$

expressed in terms of linguistic variables (such as big negative (BN), means negative (MN), small negative (SN), zero (Z), small positive (SP), means positive (MP), and big positive (BP).

We can represent the WECS controlled by FLC as in Fig. 3.31.

3.6 Adaptative Fuzzy Logic Controller

The adaptative fuzzy logic controller (AFLC) is improved from scaling FLC if any one of its tunable parameters (membership functions, fuzzy rules, and scaling factors) changes when the controller is being used, if not, it is a conventional fuzzy controller. The error (e) and the variation error (Δe) of the system and of the modifier-based learning are used to modify the fuzzy parameters to optimize system operation. The errors are given by [34]:

$$e(k) = \frac{P_{\text{wind}}(k+1) - P_{\text{wind}}(k)}{\Omega_t(k+1) - \Omega_t(k)} \tag{3.27}$$

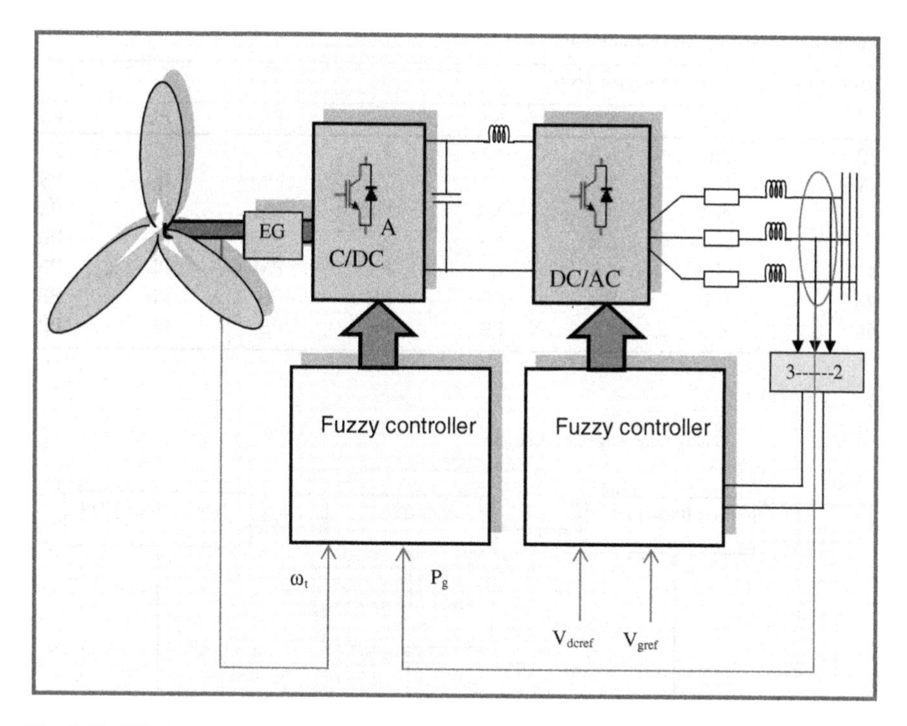

Fig. 3.31 Wind energy conversion system controlled by fuzzy logic controller

And the error variation $\Delta e(k)$ is

$$\Delta e(k) = e(k+1) - e(k) \tag{3.28}$$

The input $e(k)$ shows if the load operation point at the instant k is located on the left or on the right of the maximum power point, while the input $\Delta e(k)$ expresses the moving direction of this point. The fuzzy parameters can be adjusted using the following condition: If $e < \varepsilon$ (limit value), then the modifier-based learning will be selected. The controller Mamdani type with seven classes' membership functions is represented in Table 3.2.

The AFLC method is composed of two parts: the fuzzy logic control and adaptive mechanism. The FLC is one part of AFLC, which is composed of three units: fuzzification, fuzzy rules, and defuzzification [34] (Fig. 3.32).

Figure 3.33 shows the membership function of AFLC.

3.7 Artificial Neural Networks Method

Artificial neural networks (ANN) are electronic models based on the neural structure of the brain. This function permits ANNs to be used in the design of adaptive and intelligent systems since they are able to solve problems from

Table 3.2 Modified fuzzy rules table

Error (e)	Variation error ($\Delta\varepsilon$)						
	NB	NM	NS	Z	PS	PM	PM
NB	NB	NB	NM	Z	Z	Z	Z
NM	NB	NM	NM	Z	NM	PS	PS
NS	NB	NB	NB	NB	PM	PS	PM
Z	NB	NB	NS	Z	PS	PM	PB
PS	NM	NS	Z	PS	PM	PB	PB
PM	NS	PB	PB	PB	PB	PB	PB
PB	Z	PB	PB	PB	PB	PB	PB

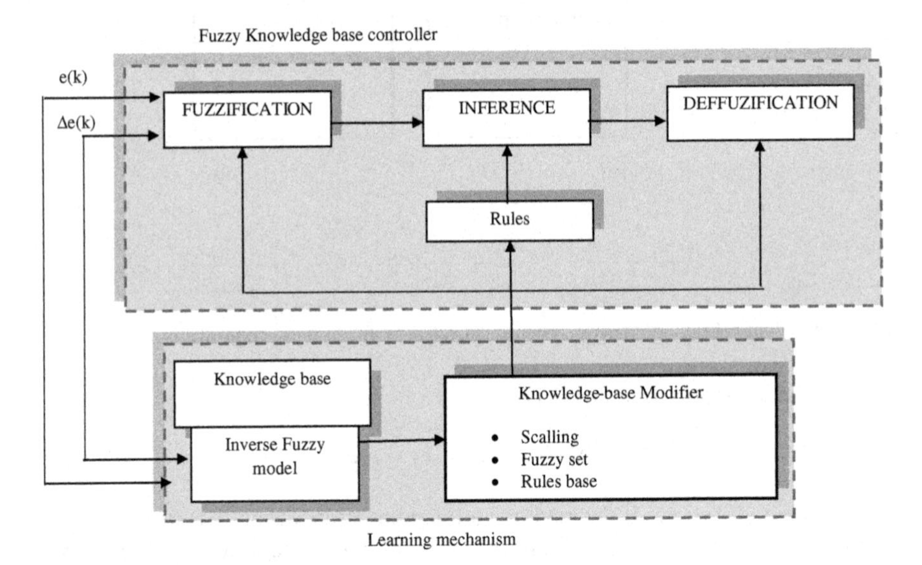

Fig. 3.32 Block diagram of adaptative fuzzy logic controller

previous examples. ANN models involve the creation of massively paralleled networks composed mostly of nonlinear elements known as neurons. Each model involves the training of the paralleled networks to solve specific problem [33]. ANNs consist of neurons in layers, where the activations of the input layer are set by an external parameter. Generally, networks contain three layers—input, hidden, and output. The input layer receives data usually from an external source, while the output layer sends information to an external device. There may be one or more hidden layers between the input and output layers. The back-propagation method is the common type of learning algorithm [33].

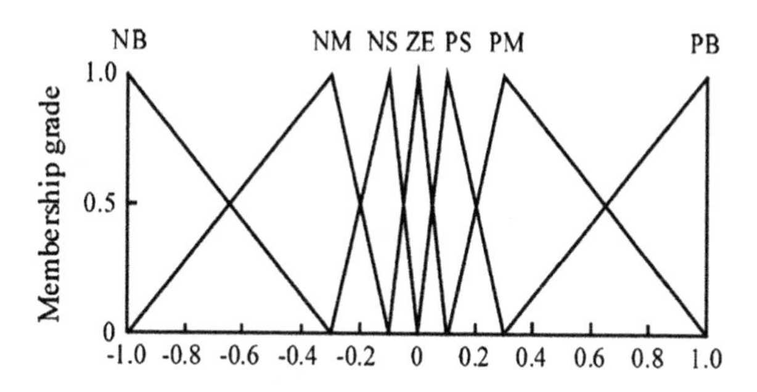

Fig. 3.33 Membership functions of AFLC method

3.8 Radial Basis Function Network

Radial basis function network (RBFN) has a similar feature to fuzzy system. The output value is calculated using the weighted sum method, and the number of nodes in the hidden layer of the RBFN is the same as the number of if–then rules in the fuzzy system. The receptive field functions of the RBFN are similar to the membership functions of the premise part in the fuzzy system. An application of RBFN on a wind energy conversion system is represented in Fig. 3.34. The electrical generator is driven by a wind turbine supplying the power to a load (grid for example), through back-to-back converters. To control the DC/AC converter, we use an MPPT (P&O and RBFN) control for maximizing power and a PWM control. The reference dc voltage V_{dcref} is obtained using P&O method and RBFN controller force V_{dc} to follow its reference V_{dcref} and adjust the load current reference I_{Loadref} for the PWM inverter control (Figs. 3.35 and 3.36).

3.9 Particle Swarm Optimization Method

Particle swarm optimization (PSO) method applies an analogy of swarm behavior of natural creatures (birds or fish), for example, the schooling of fish and the flocking of birds. Birds usually seek food (their objective) in swarms. Each individual bird (agent) reconfigures its behavior, based on its own experience and the experiences of other [30–32].

A swarm is a population of particles, and each particle flies toward the optimum or a quasi-optimum solution based on its own experience, experience of nearby particles, and global best position among particles in the swarm. At time t, each particle i has its position X_t^i and a velocity V_t^i in a variable space. The velocity and position of each particle change in the next generation (X_i^{k+1}, V_i^{k+1}) (Fig. 3.37).

Where i individual particle, X_i^k current position of the particle i, X_i^{k+1} modified position of the particle i. $P\mathrm{best}_i$ velocity based on personal best, $G\mathrm{best}_i$ velocity based on global test, V_i^k current velocity of the particle i, and V_i^{k+1} modified velocity of the particle i.

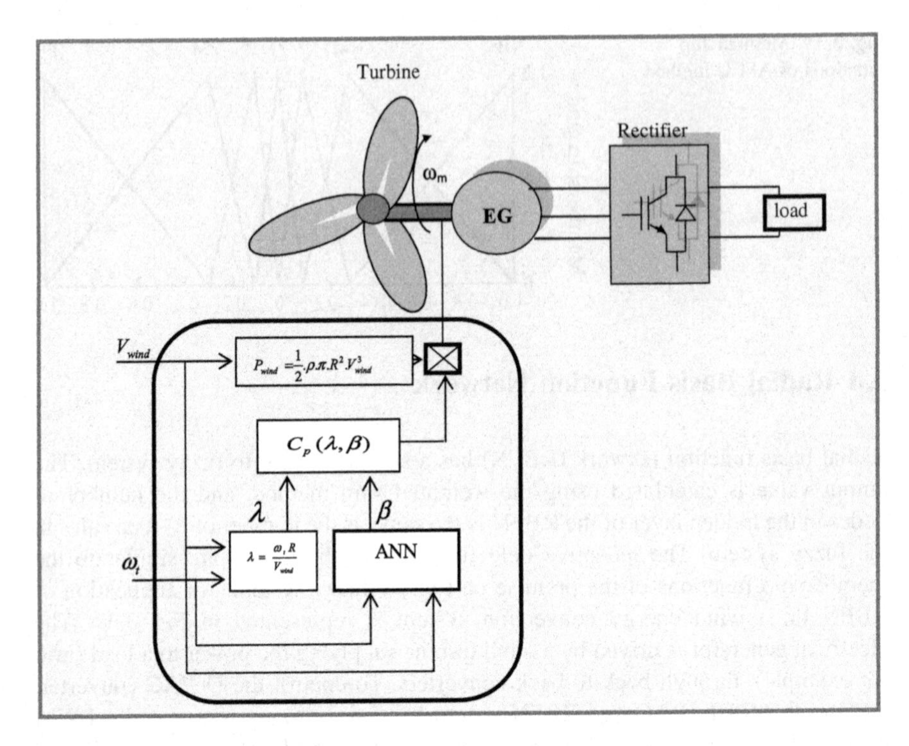

Fig. 3.34 ANN control of wind energy conversion system

3.9.1 Adaptative Neuro-Fuzzy Inference System

The adaptative neuro-fuzzy inference system (ANFIS) controller is designed and adapted to tracking a maximum power of the wind. ANFIS is the integration of artificial neural networks and fuzzy inference systems [33]. Neural network (NN) is used to adjust input and output parameters of membership function in the FLC. A typical architecture of a neuro-fuzzy network for two inputs (x and y) is shown in Fig. 3.38.

The first layer is called input layer. Each node of this layer stores the parameters to define a bell-shaped membership function. In the second layer, each node performs connective operation "AND" within the rule antecedent to determine the corresponding firing, and the nodes of layer 3 perform a normalization process to produce the normalized firing strength. The fourth layer deals with the consequent part of the fuzzy rule. The node of this layer is adaptive with output. Finally, the fifth layer which is the final output is the weighted average of all rule outputs [33].

The first of three input signals of ANFIS is the error signal $e(t)$, the second one is the changing of error signal depending on time $\mathrm{d}e(t)/\mathrm{d}t$, and the third input signal of ANFIS is the value of power P_{mec}.

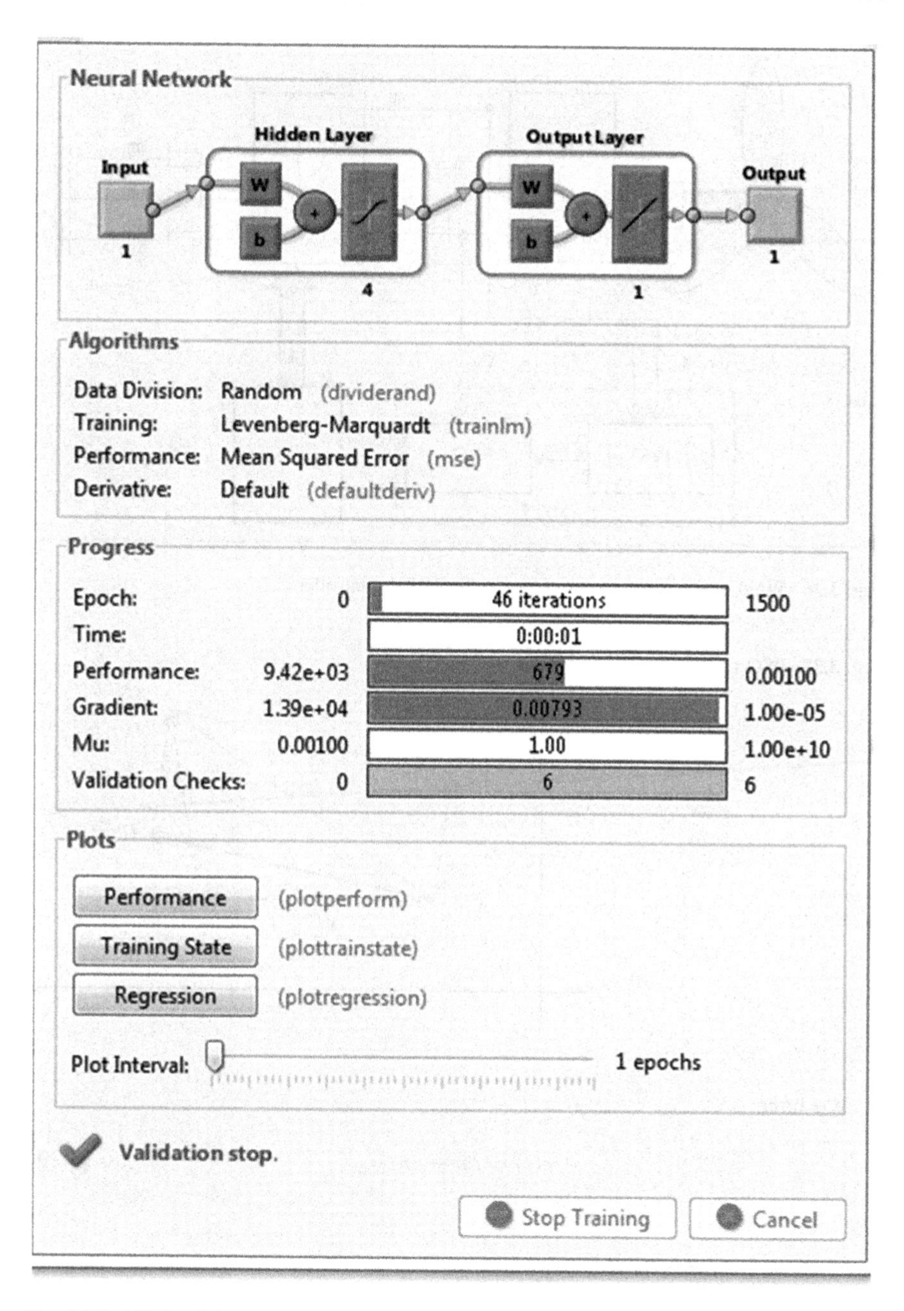

Fig. 3.35 ANN training

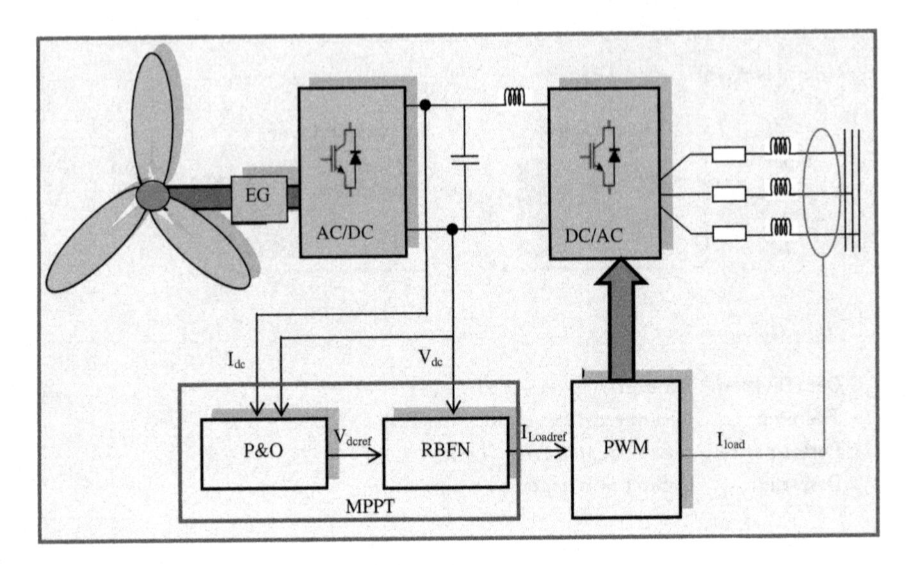

Fig. 3.36 Wind energy conversion system with RBFN controller

Fig. 3.37 PSO concept

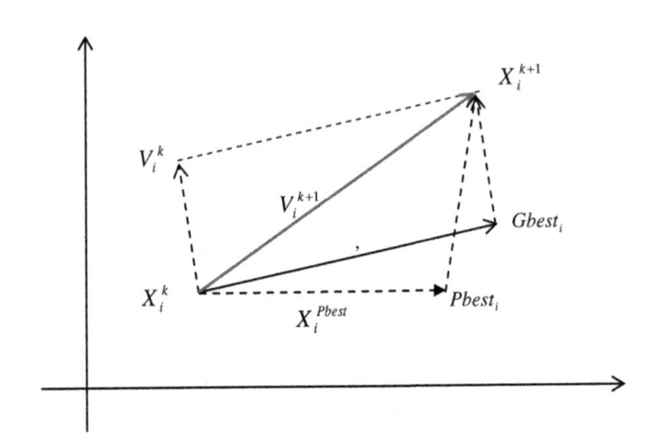

We have:

$$e(t) = \omega_t - \omega_{t-\text{nom}} \tag{3.29}$$

And:

$$\frac{\mathrm{d}e(t)}{\mathrm{d}t} = \frac{\omega_t - \omega_{t-\text{nom}}}{\mathrm{d}t} \tag{3.30}$$

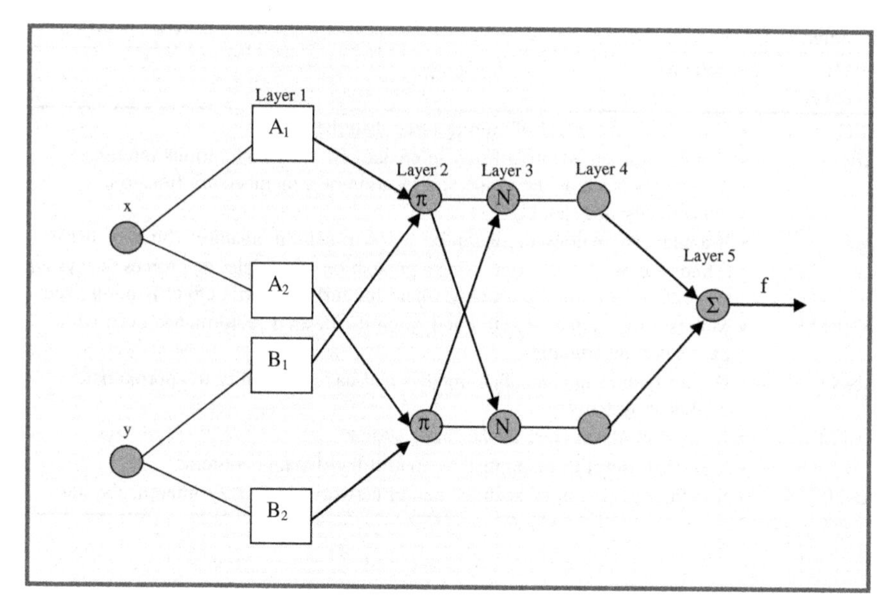

Fig. 3.38 General structure of neuro-fuzzy controller

3.9.2 Comparison Between Different Optimization Methods

MPPT methods	Proprieties
TSR	• Regulates the wind turbine rotor speed to maintain an optimal tip speed ratio • The TSR direction control method is limited by the difficulty in wind speed and turbine speed measurements
PSF	• Requires the knowledge of the wind turbine's maximum power curve and tracks this curve through its control mechanisms.
HCS or P&O	• Its purpose is to search continuously for the peak output power of the wind turbine • It is a popular method due to its simplicity and independence of system characteristics • It can be fast and effective in spite of the variations in wind speeds and the presence of turbine inertia • It is simple to implement
SMC	• High precision • Good stability • Simplicity, invariance • Robustness, etc.

(continued)

(continued)

MPPT methods	Proprieties
FLC	• Can reduce the effect of cutting force disturbances
AFLC	• Can perform an adaptive fuzzy inference process using various inference parameters, such as the shape and location of a membership function, dynamically and quickly
PSO	• It applies an analogy of swarm behavior of natural creatures (birds or fish)
	• It finds the optimal solution using a population of particles and forces the system to reach its equilibrium quickly where the turbine inertia effect is minimized
WRBFM	• Maintain the system stability and reach the desired performance even with parameter uncertainties
ANN	• Do not require mathematical models and have the ability to approximate nonlinear systems
RBFM	• It has a similar feature to the fuzzy system
	• It is very useful to be applied to control the dynamic systems
ANFIS	• It is the integration of artificial neural networks and fuzzy inference systems

References

1. Wang Q, Chang L (2004) An intelligent maximum power extraction algorithm for inverter-based variable speed wind turbine systems. IEEE Trans Power Electron 19(5):1242–1249
2. Mirecki A (2005) Phd Thesis, Institut National Polytechnique de Toulouse, France
3. Koutroulis E, Kalaitzakis K (2006) Design of a maximum power tracking system for wind-energy-conversion applications. IEEE Trans Industr Electron 53(2):486–494
4. Raza KSM, Goto H, Guo H, Ichinokura O (2008) A novel algorithm for fast and efficient maximum power point tracking of wind energy conversion systems. In: Proceedings of the 2008 international conference on electrical machines, pp 1–6
5. González LG, Figueres E, Garcerá G, Carranza O (2010) Maximum-power-point tracking with reduced mechanical stress applied to wind-energy-conversion-systems. Appl Energy 87:2304–2312
6. Soetedjo A, Lomi A, Mulayanto WP (2011) Modeling of wind energy system with MPPT control. In: International conference on electrical engineering and informatics, Bandung, Indonesia, 17–19 July 2011
7. Lin WM, Hong CM (2011) Wind Turbines, Chap. 13. InTech book
8. Serban I, Marinescu C (2012) A sensorless control method for variable-speed small wind turbines. Renewable Energy 43:256–266
9. Merabet Boulouiha H, Allali A, Tahri A, Draou A, Denaï M (2012) A simple maximum power point tracking based control strategy applied to a variable speed squirrel cage induction generator. J Renewable Sustain Energy 4(5)
10. Raza KSM, Goto H, Guo H, Ichinokura O (2010) Review and critical analysis of the research papers published till date on maximum power point tracking in wind energy conversion system. IEEE energy conversion congress and exposition (ECCE'2010), pp 4075–82
11. Wang Q, Chang L (2003) An intelligent maximum power extraction algorithm for inverter-based variable speed wind turbine systems. IEEE Trans Power Electron 19(5):1242–1249
12. Shirazi M, Viki AH, Babayi O (2009) A comparative study of maximum power extraction strategies in PMSG wind turbine system. In: IEEE electrical power & energy conference (EPEC'2009), pp 1–6

13. Morimoto S, Nakayama H, Sanada M, Takeda Y (2005) Sensorless output maximization control for variable-speed wind generation system using IPMSG. IEEE Trans Ind Appl 41(1)
14. Wai RJ, Lin CY, Chang YR (2007) Novel maximum-power extraction algorithm for PMSG wind generation system. IET Electr Power Appl 1(2):275–283
15. Femia N, Granozio D, Petrone G, Spagnuolo G, Vitelli M (2007) Predictive & adaptive MPPT perturb and observe method. IEEE Trans Aerosp Electron Syst 43(3):934–950
16. Hui J, Bakhshai A (2008) A fast and effective control algorithm for maximum power point tracking in wind energy systems. In: Proceedings of the 2008 world wind energy conference, pp 1–10
17. Molina MG, Mercado PE (2008) A new control strategy of variable speed wind turbine generator for three-phase grid-connected applications. In: Transmission and distribution conference and exposition, IEEE/PES, Bogota, Colombia, 13–15 Aug 2008
18. Lalouni S, Rekioua D, Rekioua T, Matagne E (2009) Fuzzy logic control of stand-alone photovoltaic system with battery storage. J Power Sources 193:899–907
19. Salas V, Olias E, Barrado A, Lazaro A (2006) Review of the maximum power point tracking algorithms for stand-alone photovoltaic systems. Sol Energy Mater Sol Cells 90(11):1555–1578
20. Yaoqin J, Zhongqing Y, Binggang C (2002) A new maximum power point tracking control scheme for wind generation. In: International conference on power system technology (PowerCon'2002), 13–17 Oct 2002, pp 144–148
21. Martinez-Rojas M, Sumper A, Gomis-Bellmunt O (2010) Reactive power management in wind farms using PSO technique. In: EPE J
22. Das DC, Roy AK, Sinha N (2011) PSO optimized frequency controller for wind-solar thermal-diesel hybrid energy generation system: a study. Int J Wisdom Based Comput 1(3):128–133
23. Mohandes M et al (2011) Estimation of wind speed profile using adaptive neuro-fuzzy inference system (ANFIS). Appl Energy. doi:10.1016/j.apenergy.2011.04.015
24. Dadone A, Dambrosio L (1999) Estimator-based adaptive fuzzy logic control technique for a wind turbine-induction generator system. In: Proceedings of the 7th Mediterranean conference on control and automation (MED99) Haifa, Israel, 28-30 June 1999
25. Swierczynski M, Teodorescu R, Rasmussen CN, Rodriguez P, Vikelgaard H (2010) Overview of the energy storage systems for wind power integration enhancement. In: Proceedings of the IEEE international symposium on industrial electronics, ISIE 2010. IEEE Press, p 3749–3756
26. Wang W (2012) Vanadium Redox flow batteries improving the performance and reducing the cost of vanadium redox flow batteries for large-scale energy storage. Report of Pacific Northwest national laboratory, Oct 2012
27. Comparison of different battery technologies (2006) General Electronics Battery Co. Ltd., p 1–4. www.tradekorea.com/product/file/download.mvc;…TK
28. Britton DL, Miller TB (2000) Battery fundamentals and operations-batteries for dummies, Apr 2000
29. Linden D (2002) Handbook of batteries and fuel cells, 3rd edn. Mcgraw-Hill, New York
30. Divya KC, Stergaard J (2009) Battery energy storage technology for power systems—an overview. Electr Power Syst Res 79:511–520
31. Hussien ZF, WC Lee, Siam M, Ismail AB (2007) Modeling of sodium sulfur battery for power system applications. Elektrika 9(2):66–72. (http://fke.utm.my/elektrik)
32. Survey of Energy Storage Options in Europe (2010) Report © London Research International Ltd, London, Mar 2010
33. Piemontesi M, Dustmann C (2010) Energy storage systems for ups and energy management at consumer level; (www.battcon.com/papersfinal2010/piemontesipaper2010final_5.pdf
34. Manzoni R, Metzger M, Crugnola G (2008) Zebra electric energy storage system: from R&D to market. Presented at HTE hi.tech.expo—Milan 25–28 Nov 2008

Chapter 4
Modeling of Storage Systems

4.1 Introduction

Energy storage is a dominant factor. It can reduce power fluctuations, enhance system flexibility, and enable the storage and dispatch of electricity generated by variable renewable energy sources such as wind and solar. Different storage technologies are used with wind energy system or with hybrid wind systems. It can be electrical, chemical or electrochemical, mechanical, or thermal. Energy storage facility is comprised of storage medium, power conversion system, and balance of plant.

4.2 Electrochemical Storage

4.2.1 Electrochemical Batteries

The desired battery is obtained when two or more cells are connected in an appropriate series/parallel arrangement to obtain the required operating voltage and capacity for a certain load (Fig. 4.1).

In the market, there are many different types of batteries and most of them are subject to further research and development. In PV systems, several types of batteries can be used: nickel–cadmium (Ni–Cd), nickel–zinc (Ni–Zn), lead–acid, etc. Nevertheless, it must have some important properties as high charge or discharge efficiency, low self-discharge, long life under cyclic by charging or by discharging, etc.

4.2.1.1 Nickel–Cadmium Batteries

The Ni–Cd batteries are commonly known as relatively cheap and robust. The positive nickel electrode is a nickel hydroxide/nickel oxyhydroxide ($Ni(OH)_2$/ NiOOH) compound, while the negative cadmium electrode consists of metallic

D. Rekioua, *Wind Power Electric Systems*, Green Energy and Technology,
DOI: 10.1007/978-1-4471-6425-8_4, © Springer-Verlag London 2014

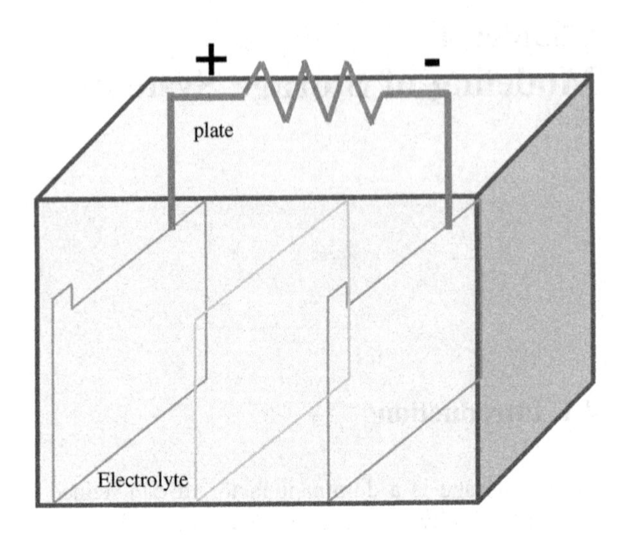

Fig. 4.1 Battery cell composition

cadmium (Cd) and cadmium hydroxide ($Cd(OH)_2$). The electrolyte is an aqueous solution of potassium hydroxide (KOH). Due to its disadvantage of life span and environmental technology of cadmium, Ni–Cd batteries are not very applicable in WECS.

4.2.1.2 Nickel–Hydrogen Batteries

Nickel–Hydrogen battery has some advantages as long cycle, resistance to overcharge, and good energy density, but it has high cost, high cell pressure, and low volumetric energy density. It is used generally in space applications and communication satellites.

4.2.1.3 Nickel–Metal Hydride Batteries

These batteries are used generally in commercial consumer product. Their disadvantages are high self-discharge and failure leading to high pressure.

4.2.1.4 Nickel–Zinc Batteries

The positive electrode is the nickel oxide, but the negative electrode is composed of zinc metal. In addition to better environmental quality, this type of battery has a high energy density (25 % higher than nickel–cadmium).

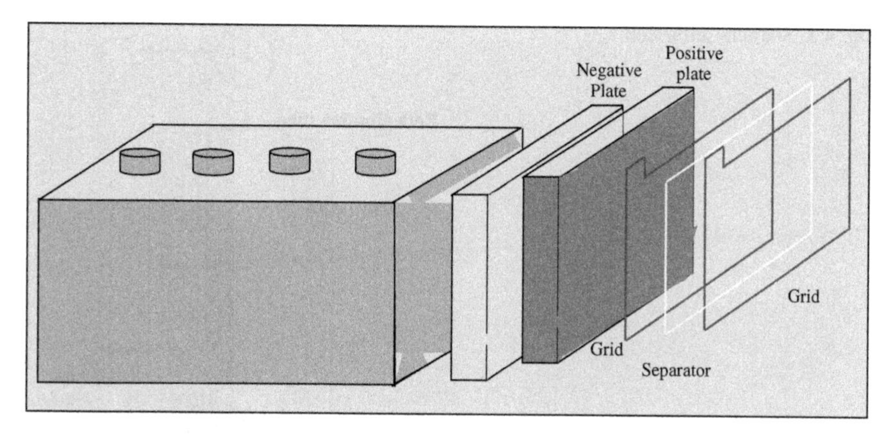

Fig. 4.2 Lead–acid battery

4.2.1.5 Lead–Acid Batteries

The lead–acid batteries are the most used in PV applications especially in stand-alone power systems because of its spill proof and the ease to transport. The lead–acid battery consists of two electrodes immersed in sulfuric acid electrolyte. The negative one is attached to a grid with sponge metallic lead, and the positive one is attached to a porous grid with granules of metallic lead dioxide.

There are two types of lead–acid batteries (Fig. 4.2):

- flooded (FLA)
- valve-regulated (VRLA).

In wind energy conversion system, electrochemical batteries are often used due to their advantages, but due to their disadvantages (smaller power density, low depth of discharge, life cycle, lifetime, etc…), research is focused on ultra-battery (see Sect. 4.5.1) which is a battery with integrated super capacitor in one unit cell providing high power discharge and charge with a long, low-cost life [1].

4.2.1.6 Sodium–Sulfur Batteries

In a sodium–sulfur (NaS) battery, sodium and sulfur are in liquid form and are the electrodes, sodium being the cathode and sulfur being the anode. They are separated by alumina which has the role of electrolyte. This one allows only the positive sodium ions to go through it and combine with the sulfur to form sodium polysulfide. This type of battery has a high energy density, high efficiency of charge/discharge (89–92 %), and long life cycle, and it is fabricated from inexpensive materials (Fig. 4.3).

Due to their advantages (high energy density, high number of cycles), Nas batteries are very used in WECS.

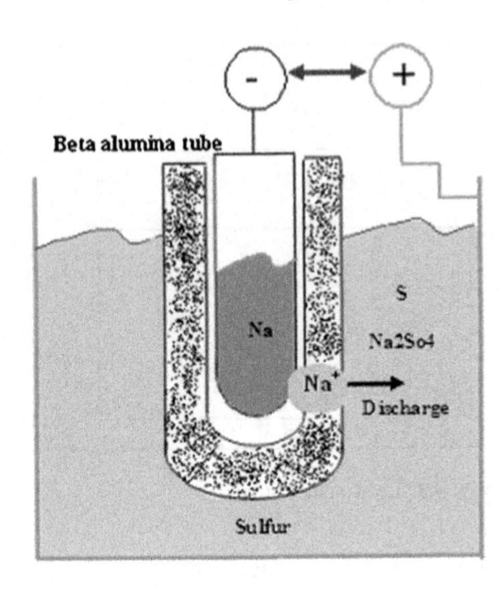

Fig. 4.3 Sodium–sulfur cell

4.2.1.7 Sodium–Nickel Chloride Batteries

Sodium–nickel chloride battery is also known as Zero Emission Battery Research Activity (ZEBRA) battery, and it is a system operating at around 270–350 °C. The chemical reaction in the battery converts sodium chloride and nickel to nickel chloride and sodium during the charging phase. During discharge, the reaction is reversed. Each cell is enclosed in a robust steel case. A ZEBRA battery is designed for a 2 h discharge with peak power capability as required.

They are used in WECS because of their advantages as high energy density and resistant for short circuits.

4.2.1.8 Lithium-ion Batteries

The operation of Lithium-ion (Li-ion) batteries is based on the transfer of lithium ions from the positive electrode to the negative electrode during charging and vice versa during discharging. The positive electrode of a Li-ion battery consists of one of a number of lithium metal oxides, which can store lithium ions, and the negative electrode of a Li-ion battery is a carbon electrode. The electrolyte is made up of lithium salts dissolved in organic carbonates.

Lithium-ion batteries are very used in WECS due to their advantages as small self-discharge and long life for deep cycles.

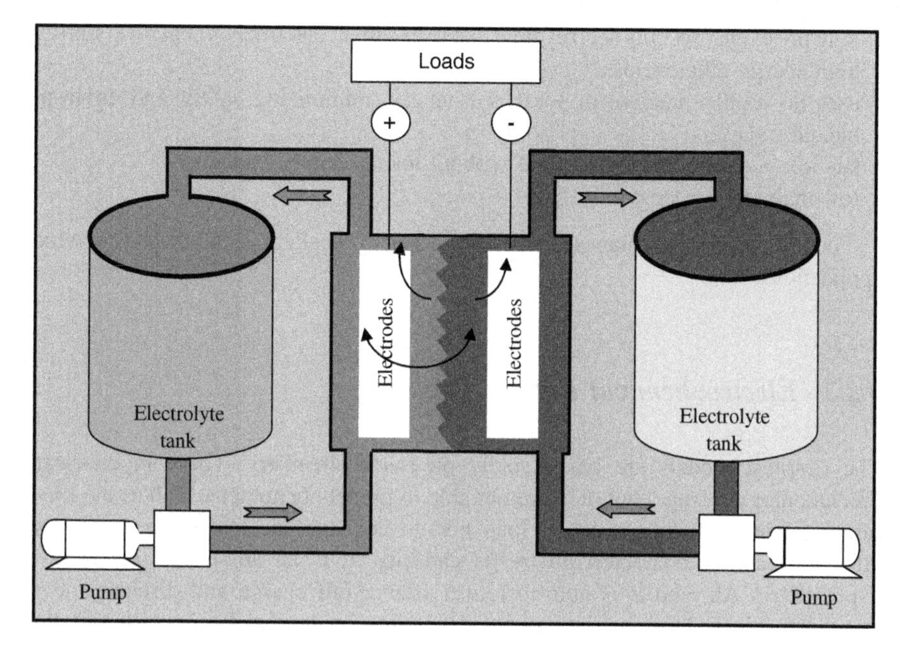

Fig. 4.4 Flow battery process

4.2.1.9 Flow Batteries or Vanadium Redox Flow Battery

The vanadium redox flow battery (VRB) stores energy in two tanks that are separated from the cell stack.

There are three kinds of flow batteries:

- Vanadium Redox (VR),
- Polysulphide Bromide (PSB),
- Zinc Bromine (ZnBr).

The flow battery process is explained in Fig. 4.4.

In flow batteries, energy is stored as a potential chemical energy and it is stored in the electrolyte solutions. The advantages of VRB are [2]:

- increasing energy densities by more than 70 % due to increased vanadium ion concentrations
- operation at increased current densities
- increasing the operating temperature window
- storing of megawatts/megawatt-hours of power and energy in simple designs
- flexibility to design power and energy capacities separately
- discharging power for up to 12 h at a time
- quickly brought up to full power when needed

- long life cycle (>5,000 deep cycles) due to excellent electrochemical reversibility
- high energy efficiencies
- uses no highly reactive or toxic substances, minimizing safety and environmental issues
- sits idle for long periods of time without losing storage capacity
- low maintenance cost.

For all these advantages, the VRB is an excellent candidate for wind applications.

4.2.2 Electrochemical Battery Model

The simplest models are based solely on electrochemistry. These models can predict energy storage, but they are not able to model phenomena such as the time rate of change of voltage under load nor do they include temperature and age effects. A cell is characterized by its capacity. It is an amount of electricity, expressed in Ah, and it is able to return after a full charge and discharge at a constant current. This capacity varies depending on several factors, such as the intensity of discharge, temperature, and electrolyte concentration.

The Peukert equation (Eq. 1.19) is an empirical formula which approximates how the available capacity of a battery changes according to the rate of discharge [2–20].

$$I_{\text{batt}}^{n} \cdot t = C \tag{4.1}$$

where I_{batt} the discharge current, n the Peukert constant.

The Peukert constant increases with age for any of the battery types (Table 4.1).

t is the time to discharge at current I_{batt}, C is the capacity according to Peukert, at a one-ampere discharge rate, expressed in Ah.

Equation (4.1) shows that at higher currents, there is less available capacity in the battery. The Peukert constant is directly related to the internal resistance of the battery and indicates how well a battery performs under continuous heavy currents. We can relate the discharge current at one discharge rate to another combination of current and discharge rate. Then, we obtain [20–31]:

$$C_1 = C_2 \cdot \left(\frac{I_{\text{batt}\,2}}{I_{\text{batt}\,1}}\right)^{n-1} \tag{4.2}$$

where C_1 and C_2 are capacities of the battery at different discharge-rate states.

Table 4.1 Peukert constant [11]

AGM batteries	Gel batteries	Flooded batteries	Typical lead–acid batteries	Lithium-ion batteries
1.05–1.15	1.1–1.25	1.2–1.6	1.35	1.1

The state of charge (SOC) at a constant discharge rate can be obtained by the following equation:

$$\mathrm{SOC}(t) = 1 - \left(\frac{I_{\mathrm{batt}}}{C}\right) \cdot t \tag{4.3}$$

The current is continuously variable over time. We discredit then the above equation by considering the constant current between two calculation steps. We can determine the expression of the change in charge state of the cell at time t_k:

$$\Delta\mathrm{SOC}(t_k) = \frac{I_{\mathrm{batt}\,k}}{C_1} \cdot \left(\frac{I_{\mathrm{batt}\,k}}{I_{\mathrm{batt}\,1}}\right)^{n-1} \cdot \Delta t \tag{4.4}$$

This approach also takes into account the phases of recharging the battery. Indeed, if the current in the cell becomes negative, its state of charge increases. Ultimately, cell state of charge expressed by:

$$\mathrm{SOC}(t_k) = \mathrm{SOC}(t_{k-1}) + \Delta\mathrm{SOC}(t_k) \tag{4.5}$$

These models are modeling the batteries in the shape of electronic circuits. There are many models proposed by different scientific research.

An application is made under Matlab/Similink, using CIEMAT model (Fig. 4.5). Some simulation results are presented (Fig. 4.6).

4.3 Hydrogen Energy Storage

Generally, hydrogen system consists of an electrolyzer, a pressurized gas store and fuel cells. The electrolyzer converts electrical energy into chemical energy in the form of hydrogen during times of surplus electrical supply. This hydrogen is stored until there is a shortage of electrical energy to power the loads on the system, and then, it is reconverted by a fuel cell (hydrogen and air oxygen) to electricity. Hydrogen can store energy for long periods. Different hydrogen storage modes are used:

- compressed,
- liquefied,
- metal hydride,
- etc.

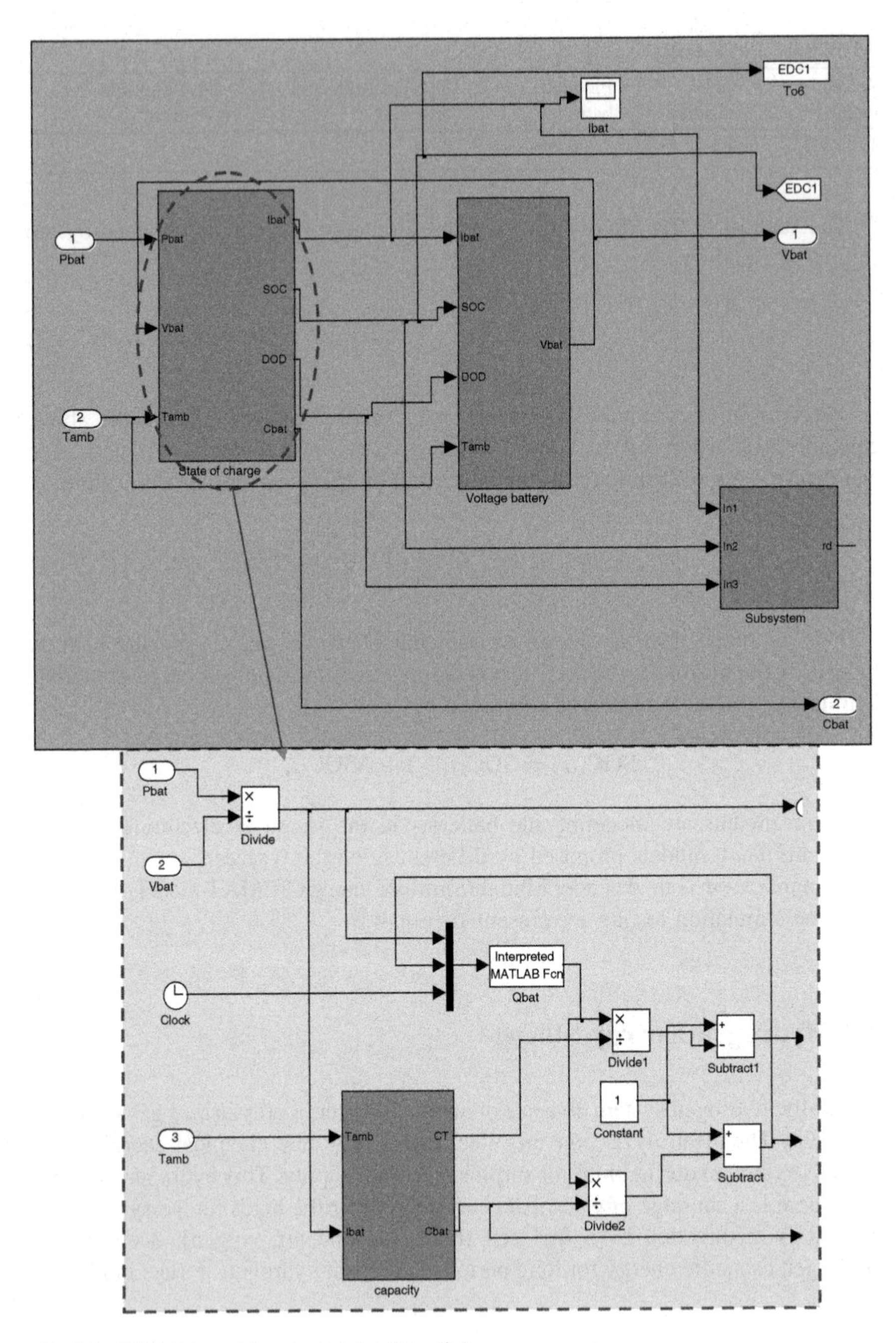

Fig. 4.5 CIEMAT model under Matlab/Simulink

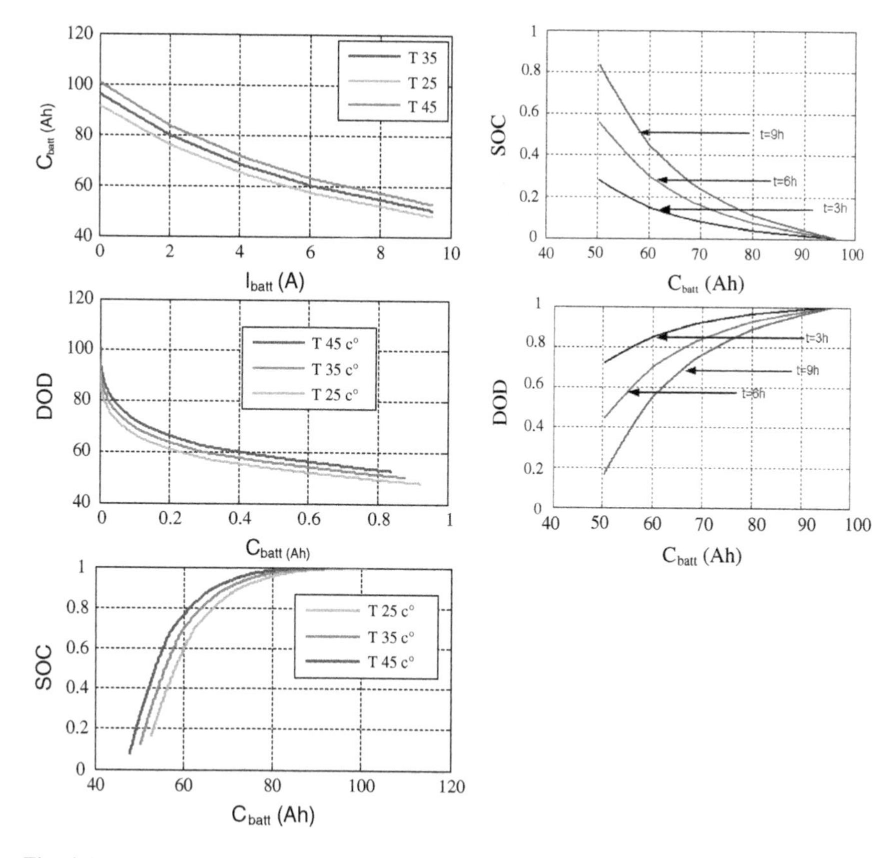

Fig. 4.6 Simulation results

For example, for a wind system or an hybrid wind/photovoltaic (or hydro) system supplying a load (Fig. 4.7), we can add battery system for short-term storage and also to stabilize the system against fluctuations in energy sources, but for a long-term storage, an electrolyzer coupled to a hydrogen storage tank is used.

The system management operates as (Fig. 4.8):

- When the energy demand loads are less than the production of wind and solar panels, the excess energy is sent to the electrolyzer to produce the hydrogen and then store it.
- When energy demand exceeds energy load capacity available, the stored hydrogen is regenerated into electricity via cell fuel.

Different Wind or hybrid system structures with hydrogen storage are proposed in scientist research and some of them are real implanted systems. In wind energy conversion system, hydrogen energy storage (HES) with all advantages (higher energy density and lower per volume than a gasoline,…) is one of the best storage solutions to suppress fast wind power fluctuations (Figs. 4.9, 4.10).

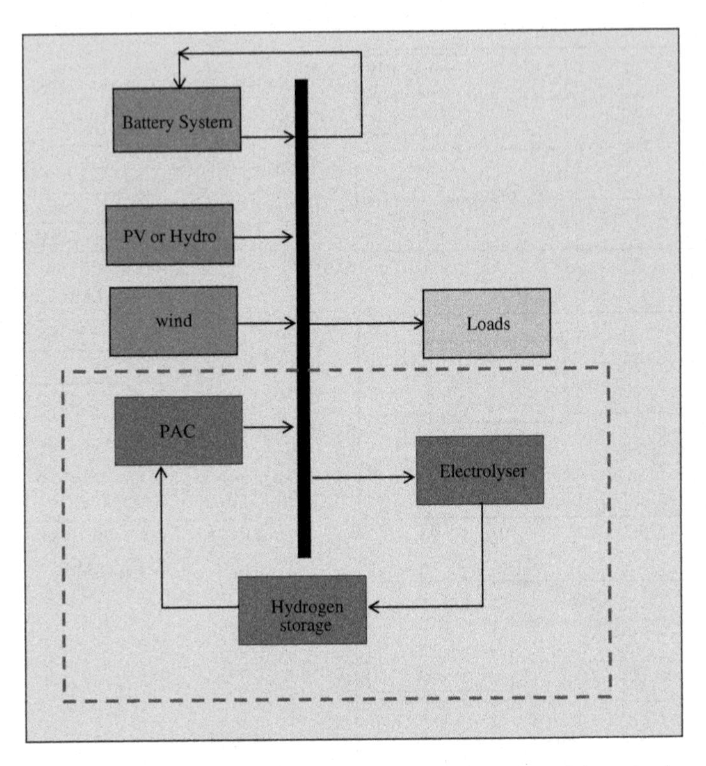

Fig. 4.7 Hybrid wind/photovoltaic system with hydrogen storage supplying a load

4.4 Mechanical Storage

4.4.1 Flywheel Energy Storage

Flywheel electric energy storage system includes a cylinder with a shaft connected to an electrical generator. Electric energy is converted by the generator to kinetic energy which is stored by increasing the flywheel's rotational speed. The stored energy is converted to electric energy via the generator, slowing the flywheel's rotational speed.

For wind standalone applications, storage cost still represents the major economic restraint. Energy storage in wind systems can be achieved in different ways. However, the inertial energy storage adapts well to sudden changes of the power from the wind generator. Moreover, it allows obtaining high power to weight and number of charge cycles and discharge very high.

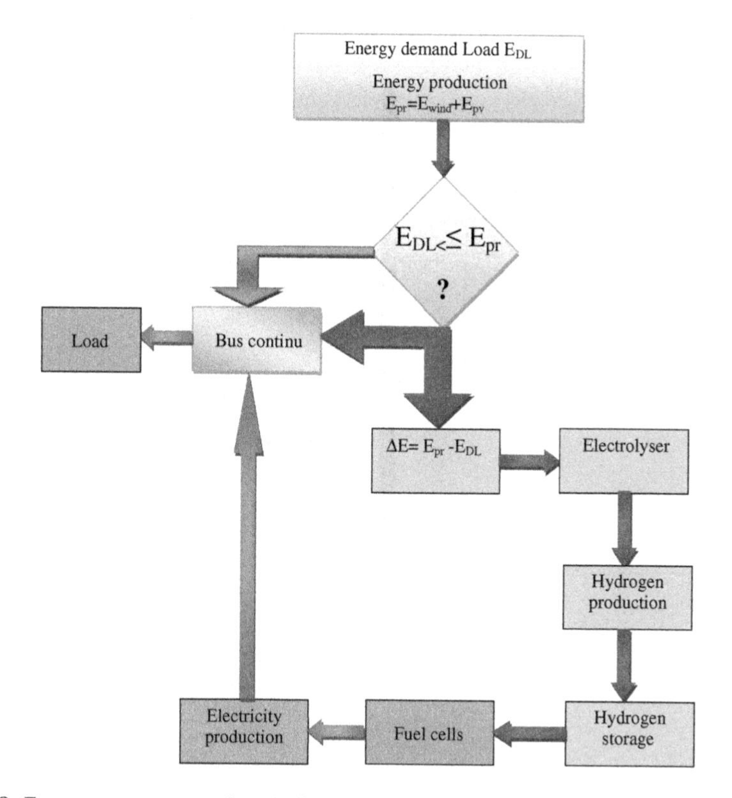

Fig. 4.8 Energy management of a wind/PV system with hydrogen storage

The reference speed for the flywheel is determined by:

$$\Omega_{\mathrm{ref}} = \sqrt{\frac{2 \cdot E_{c\,\mathrm{ref}}}{J_t}} \tag{4.6}$$

with:

$$J_t = J_{\mathrm{IG}} + J_{\mathrm{Flywheel}} \tag{4.7}$$

The reference speed is limited in order to maintain the IG in the area of operation at constant power and not exceed the maximum speed of the flywheel.

Figure 4.11 represents the torque and power as a function of speed. We notice that:

- For $0 \leq \Omega \leq \Omega_{\mathrm{rated}}$, the torque may be maximal giving up a power proportional to the speed $P_{\mathrm{IG}} = k \cdot \Omega$.
- For $\Omega \rangle \Omega_{\mathrm{rated}}$, the power is maximum and corresponds to the rated power of the machine, and the electromagnetic torque is inversely proportional to the speed $T_{\mathrm{em}} = \frac{k}{\Omega}$.

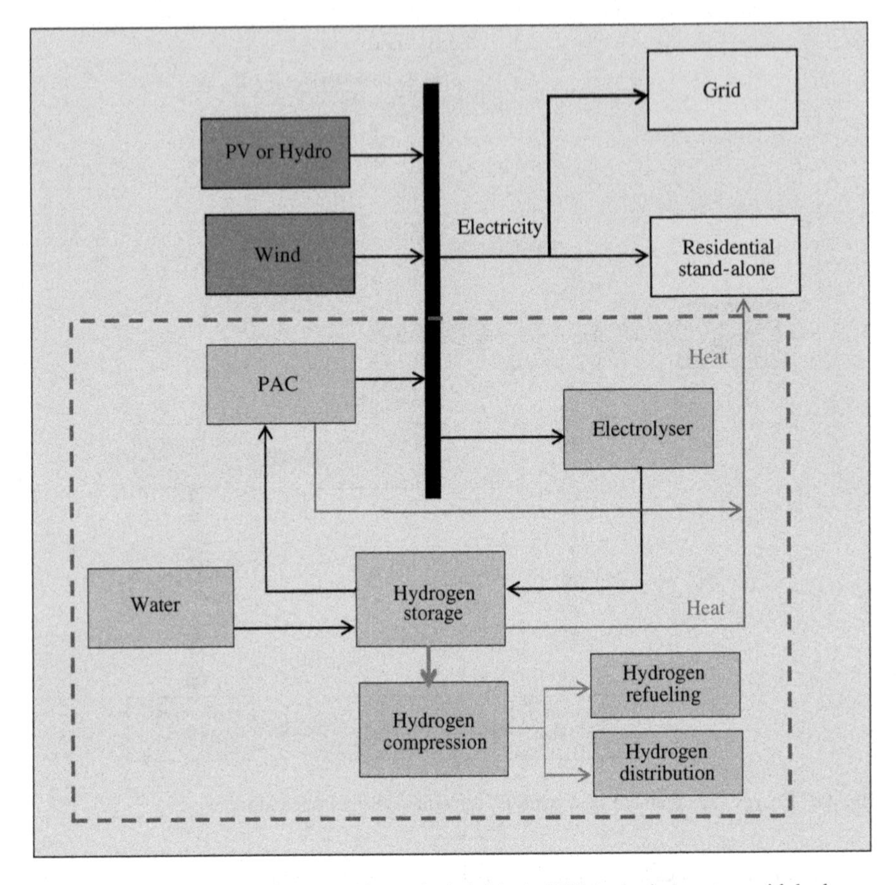

Fig. 4.9 Production of electricity and heat of a hybrid wind/PV (or hydro) system with hydrogen storage

So, if we want to have the machine rated power, it is necessary to use it beyond its rated speed, which lets us to consider the speed as the lower limit storage and the dual value of speed as the upper limit storage.

Thus, a field-weakening operation will be necessary to obtain a constant power in the speed range 1,500–3,000 rpm.

The reference flux is then determinate by:

$$\Phi_{ref} = \begin{cases} \Phi_{rated} & \Rightarrow \text{if } |\Omega| \leq \Omega_{rated} \\ \Phi_{rated} \cdot \frac{\Omega_{rated}}{|\Omega|} & \Rightarrow \text{if } |\Omega| \rangle \, \Omega_{rated} \end{cases} \tag{4.8}$$

with Ω flywheel speed, Ω_{rated}: rated speed, Φ_{rated}: rated flux, and Φ_{ref}: reference flux.

In wind energy conversion system, flywheel energy storage (FES) is able to suppress fast wind power fluctuations. We make an application to a WECS based

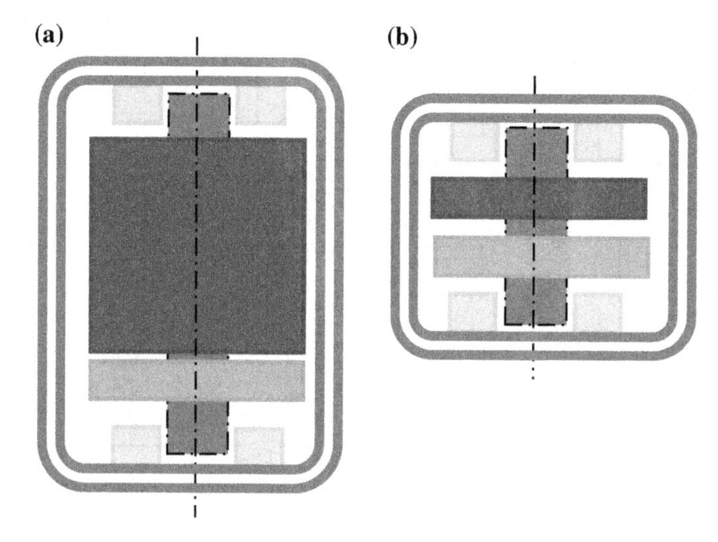

Fig. 4.10 Flywheel system **a** For long term (more than 1 h). **b** For short term (less than 1 min)

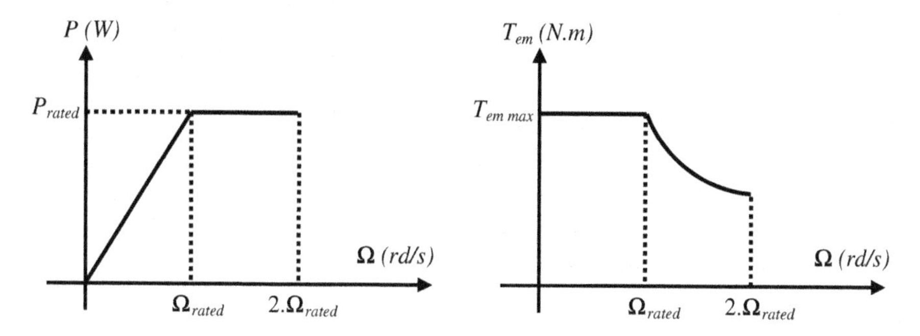

Fig. 4.11 Power and torque as a function of speed

on induction generator. The system is constituted of a wind turbine, an induction generator, a rectifier/inverter, and FES system (Fig. 4.12).

The goal of the device is to provide a constant power and voltage to the load connected to the rectifier/inverter even if the speed varies. This can be achieved mainly by the control of the DC bus voltage at a constant value, and the FES system participates to maintain the power of the load constant as long as the wind power is sufficient. To control the speed of the FES system, we must find reference speed with which the system must turn to ensure the energy transfer required at each time. The reference speed can be determinate by the reference energy. The power assessment of the overall system is given by [32]:

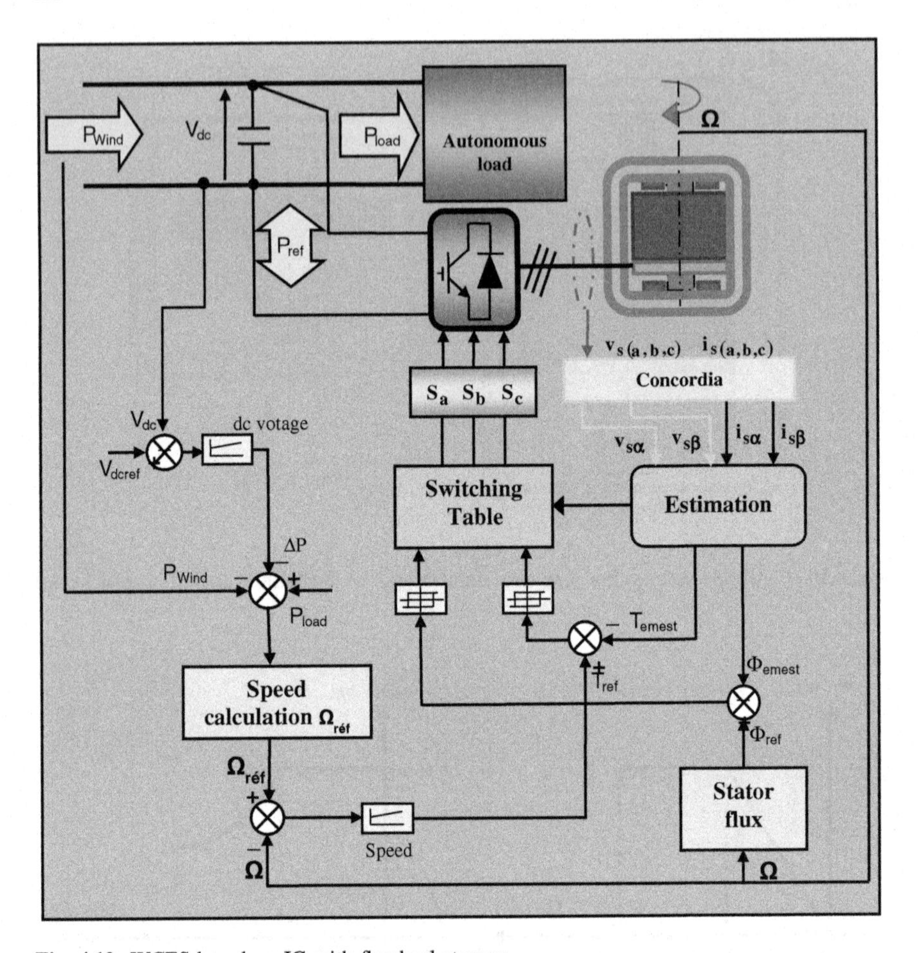

Fig. 4.12 WCES based on IG with flywheel storage

$$P_{\mathrm{ref}} = Pl_{\mathrm{oad}} - P_{\mathrm{wind}} - \Delta P \tag{4.9}$$

where P_{ref} is the reference power, P_{load} the load power, P_{wind} the wind power, and ΔP is the power required to control the DC voltage V_{dc} at constant value.

The application is made under Matlab/Simulink under a wind power profile which provides power continuously required by the load through the SISE. Figure 4.13 shows the development of wind power which varies between 1,000 and 2,400 W. We note that the power supplied to the load is kept constant through the FES system. Figure 4.14 corresponds to the storage power FES system. This power can be positive or negative. It depends on the wind power and the power load required. We note that it is positive when the wind power produced is greater than the load power required by the load and is negative when there is a less power produced compared with that of the load. Flywheel speed and the reference speed are represented in Fig. 4.15 [34].

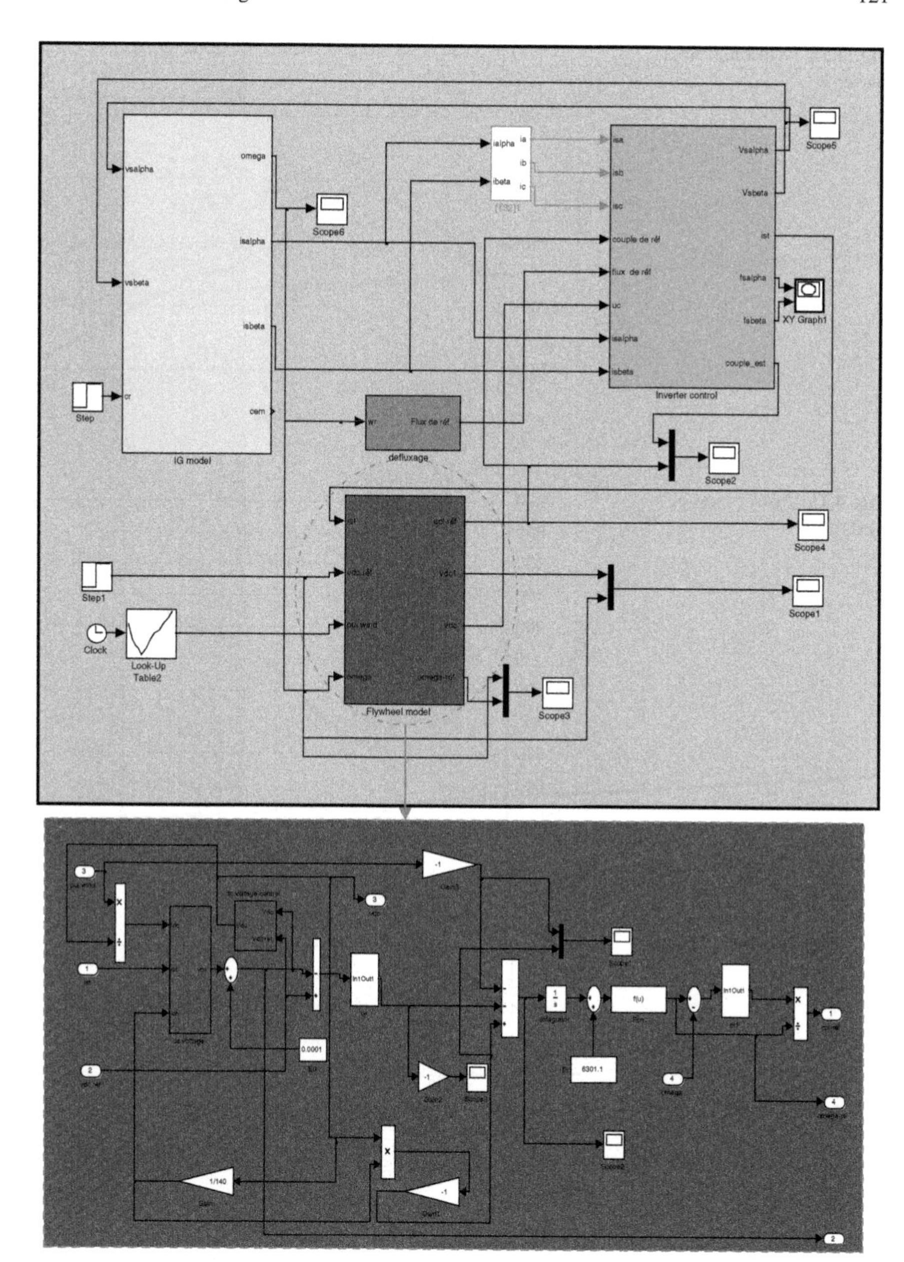

Fig. 4.13 Flywheel storage model under MATLAB/Simulink

The rotational speed increases when the energy is transferred to the flywheel and decreases when the flywheel is unloaded. The electromagnetic torque follows the evolution of the speed (Fig. 4.16). To regulate the bus voltage, we need a

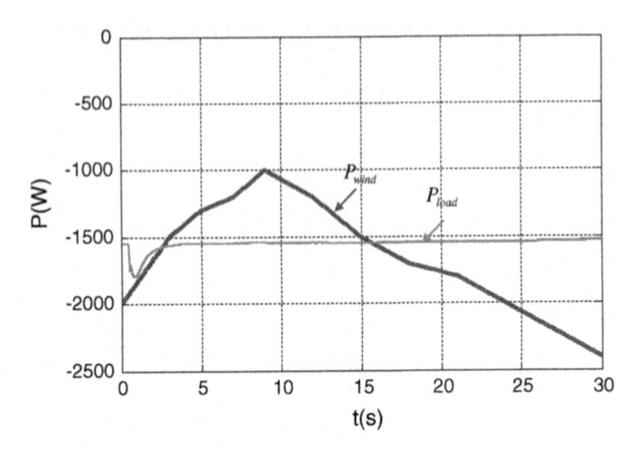

Fig. 4.14 Wind and load power

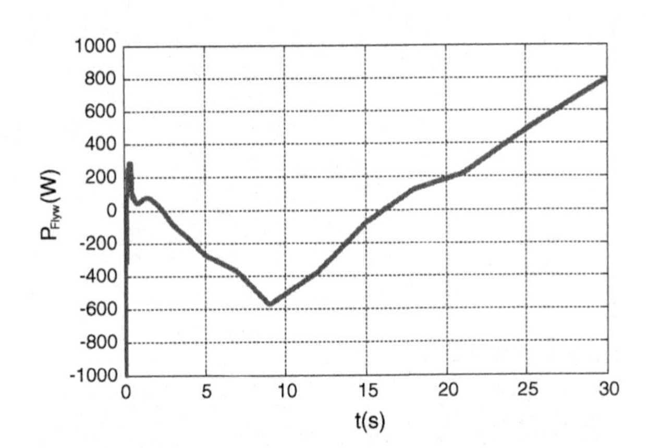

Fig. 4.15 Power storage SISE

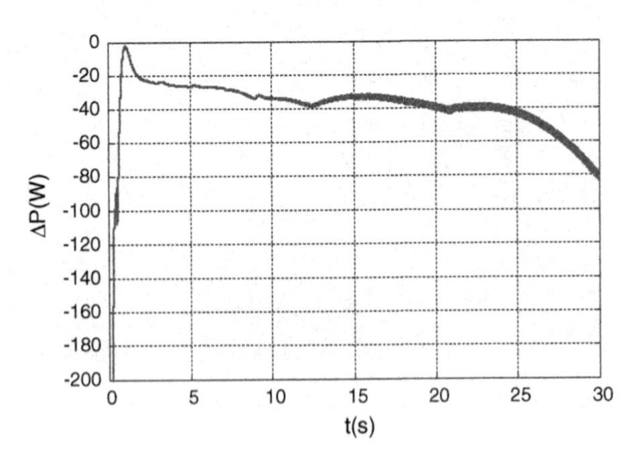

Fig. 4.16 Power of the DC bus

required power represented in Fig. 4.17. The voltage DC bus is kept constant around 465 V, with an overshoot at startup by about 10 % (Fig. 4.18). Variations of stator flux are represented in Fig. 4.19. The flux $\Phi_{s\alpha}$, and $\Phi_{s\beta}$ follows the

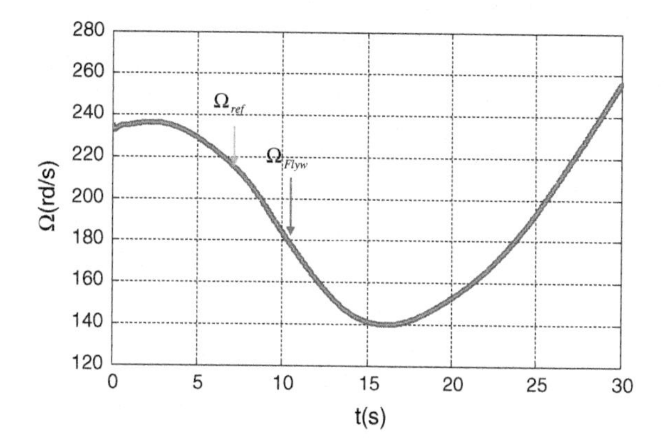

Fig. 4.17 Flywheel and reference speed

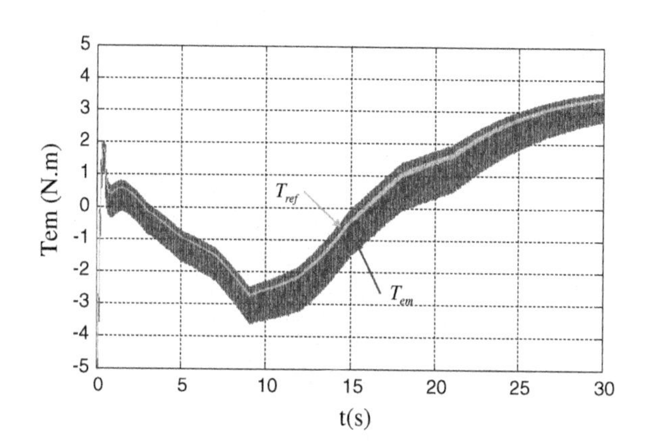

Fig. 4.18 Electromagnetic torque

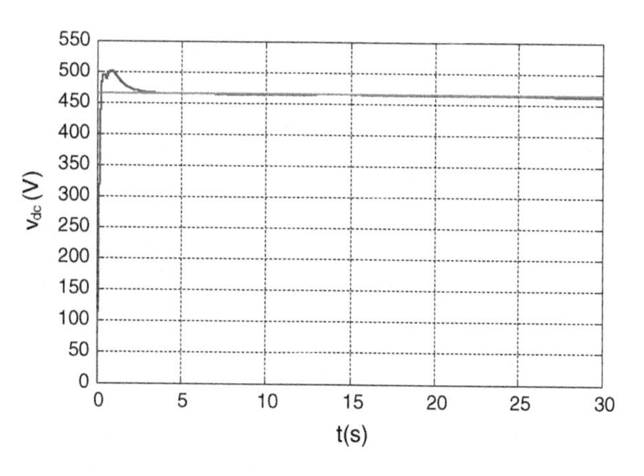

Fig. 4.19 DC voltage

variations of the speed and does not exceed the nominal flux. The results of simulation clearly show the good operation of the FES system storage (Fig. 4.20).

The simulation stator flux trajectory is represented in Figs. 4.21 and 4.22.

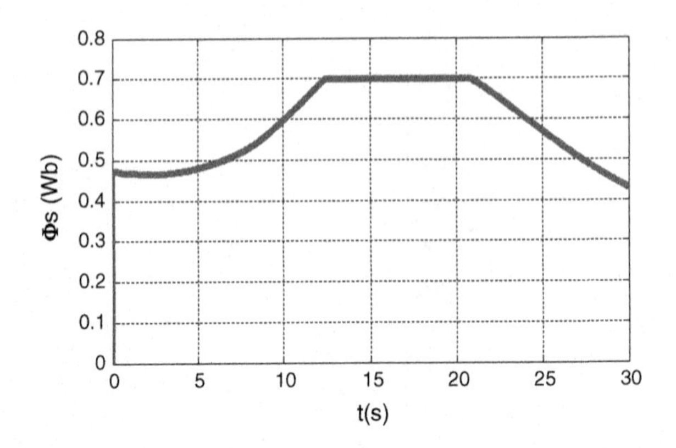

Fig. 4.20 Stator flux

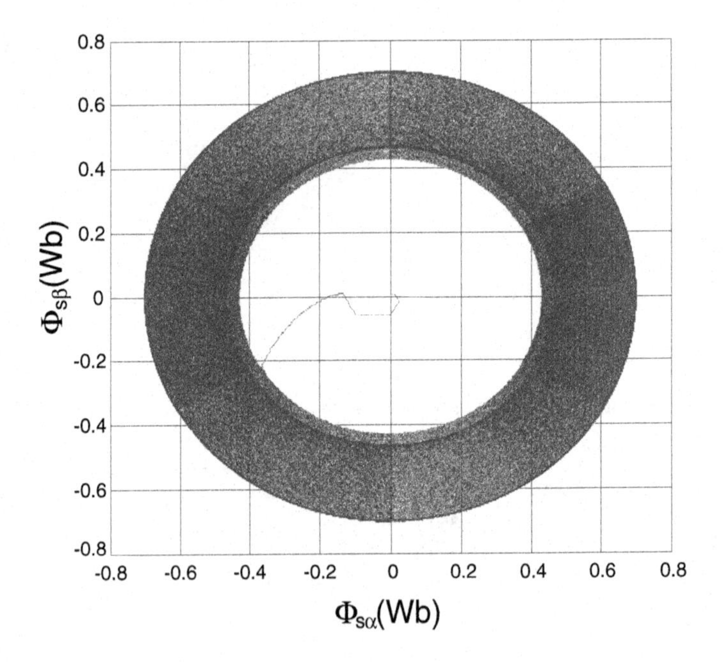

Fig. 4.21 Stator flux trajectory

4.4.2 Pumped Hydro Energy Storage

Pumped Hydro Energy Storage (PHES) system consists of a pumped hydro system with two large water reservoirs (upper and lower), electric machine (motor/generator), and reversible pump-turbine group (Fig. 4.23).

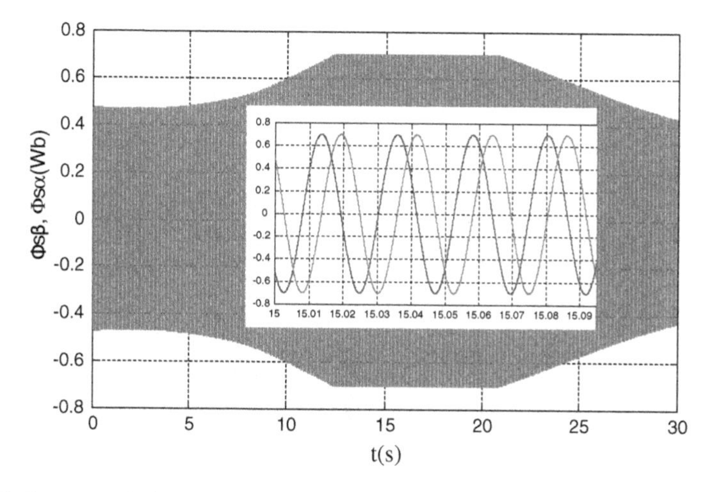

Fig. 4.22 Statot flux $\Phi s\beta$ and $\Phi s\alpha$ in real time

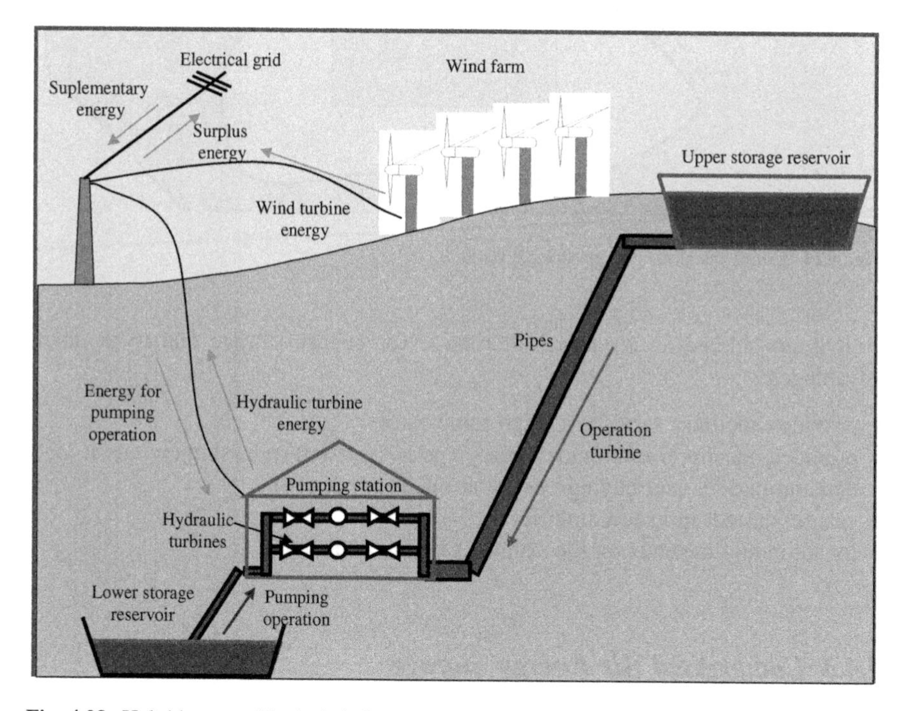

Fig. 4.23 Hybrid pumped hydro/wind energy

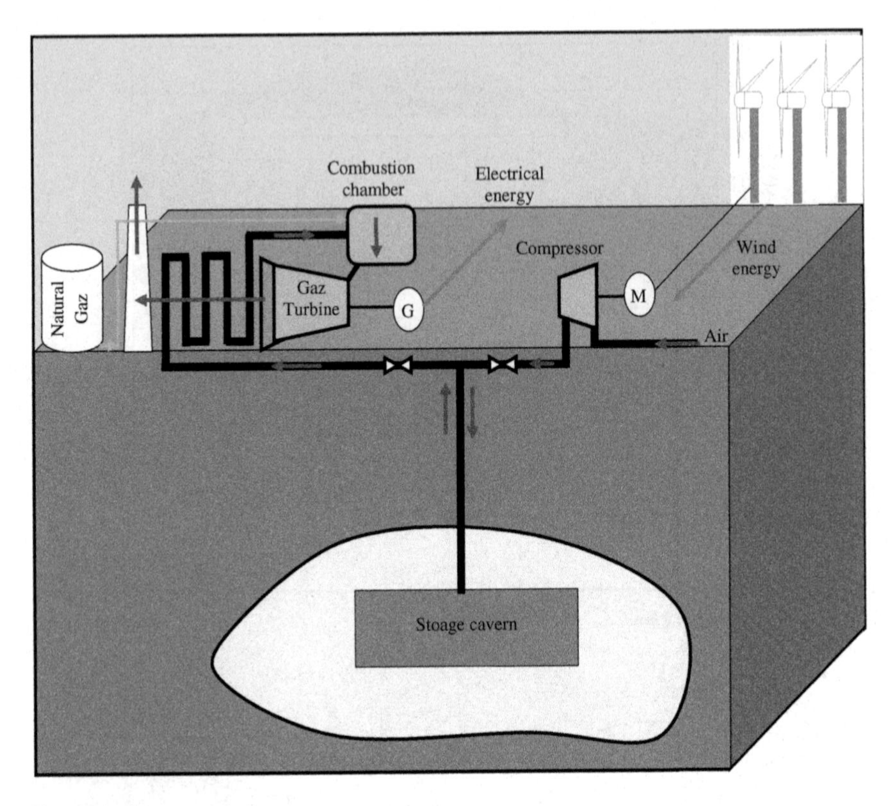

Fig. 4.24 Compressed air energy storage basis

It is considered an attractive alternative for energy storage due to its main advantages:

- provides ancillary services at high ramp rates,
- provides benefits from intraday energy price variation (releasing energy at high demand periods and buying energy at off-peak periods).
- can be started up in few minutes
- its autonomy depends on the volume of stored water.

4.4.3 Compressed Air Energy Storage

The basic idea of compressed air energy storage (CAES) is to compress air using inexpensive energy, and in its turn, the compressed air (released into a combustion turbine generator system and sent through the system's turbine) is used to generate energy. There are two types of storage:

- compressed air is stored in underground geologic formations (salt formations, aquifers) for larger CAES plants,
- compressed air is stored in tanks or large on-site pipes for smaller CAES plants.

The proposed system in Fig. 4.24 is able to provide large energy storage.

In offshore wind systems, we use pipelines as an alternative storage for compressed air. CAES is good storage solution for WECS onshore or in offshore.

4.5 Electromagnetic Storage

4.5.1 Supercapacitor Energy Storage

It is known as electric double-layer capacitors, as super capacitors (SC), electrochemical double-layer capacitors (EDLCs), or ultra-capacitors. They use polarized liquid layers between conducting ionic electrolyte and conducting electrode to increase the capacitance. They allow a much higher energy density, with a high power density, but the voltage varies with the energy stored and it has a highest dielectric absorption. Their important parameter is the relatively low, state-of-charge-dependent maximum voltage of 2.5 V and a great efficiency (around 95 %).

In wind energy conversion system, supercapacitor energy storage (SES) is used to suppress fast wind power fluctuations, but with a small timescale, thus they can be considered only as a support for wind turbine systems, and generally they are combined with a battery system as an hybrid storage system.

4.5.2 Superconducting Magnetic Energy Storage

This system consists essentially of a coil of cryogenically cooled with a superconducting material, a power conditioning system and a refrigeration system. Energy is stored in the magnetic field created by the flow of direct current in the coil. This one can be stored as long as the refrigeration is operational.

The main advantage of this system is its great efficiency and can be applied to systems requiring continuous operation and a large number of complete cycles of discharge load.

In wind energy conversion system, superconducting magnetic energy storage (SMES) is generally not used due to the coil which is very sensitive for temperature changes.

Table 4.2 Most storage technologies used in wind energy conversion system

Energy storage (ES)	Technologies			Timescale	Applications in WECS	Efficiency [33]
Electrochemical (ElES)	Batteries (BS)	Nickel–cadmium storage (NCS)		Medium (min)	X	60–70
		Nickel–hydrogen storage (NHS)				
		Nickel–metal hydride				
		Nickel–zinc				
		Sodium–sulfur storage (NaSS)				86–89
		Sodium–nickel chloride				
		Lithium-ion storage (LIS)				90–95
		Zebra				90
		Lead–acid storage (LAS)	Flooded (FLA)			75–85
			Valve-regulated (VRLA)			75–85
	Flow batteries storage (FBS)	Vanadium redox storage (VRS)		Medium (h)	X	70–80
		Polysulphide bromide storage (PSBS)				75
		Zinc bromine storage (ZnBrS)				75–80
Hydrogen (HES)	Hydrogen (HES)			Long	X	65–75
Mechanical (MES)	Flywheel energy storage (FES)			Short (s)	X	80–90
	Pumped hydro energy storage (PHES)			Long (h)	X	70–85
	Compressed air energy storage (CAES)			Long (h)	X	64–75
Electromagnetic (ElmES)	Supercapacitor energy storage (SES)			Short (s)	X	90-98
	Superconducting magnetic energy storage (SMES)			Short (s)		90–99
Thermal (TES)	Thermal (TES)			Medium	X	80–90

4.6 Thermal Energy Storage

In thermal energy storage system, we use materials that kept at high /low temperature in enclosures. We use after a heat engine to produce electrical energy, which will be powered by heat/cold recovered. Energy input can be provided by electrical resistance heating or by refrigeration/cryogenic procedures. The main applications of thermal energy storage (TES) system are [33]:

- industrial cooling (below −18 °C)
- building cooling (at 0–12 °C),
- building heating (at 25–50 °C)
- industrial heat storage (higher than 175 °C).

4.7 Conclusion

There are various energy storage systems. Each one of them has its proper proprieties as lifetime, costs, density, and efficiency. Generally, for applications, we can summarize as:

- For energy management applications, we use PHS, CAES, electrochemical batteries, flow batteries, fuel cells, solar fuels, and TES.
- For power quality and short duration, we use flywheels, batteries, capacitors, and super capacitors.
- Others applications, we use batteries, flow batteries, fuel cells, and Metal–Air cells.

We summarized in Table 4.2 the most storage technologies used in wind energy conversion system.

References

1. Rouault H, Blach D (2004) The electricity storage: the batteries, vol 44, Clefs CEA
2. Saisset R (2004) Study contribution systemic including electrochemical energy devices. Ph.D. thesis, INP Toulouse Specialty Electrical Engineering
3. Sabonnadière JC (2007) New energy technologies and storage technology decarbonizing. Edition Dunod, Paris
4. Bergveld HJ (2001) Battery management systems design by modeling. Thesis University Press Facilities, Eindhoven (doc.utwente.nl/41435/)
5. Lalouni S (2005) Optimizing the quality of electrical energy in the case of a battery charger. Thesis of master, University of Bejaia
6. Hladik J (1977) Storage batteries, first edition: 1st quarter. Presses Universities France
7. Benyahia N, Rekioua T, Benamrouche N, Bousbaine A (2013) Fuel cell emulator for supercapacitor energy storage applications. Electr Power Compon Sys 41(6):569–585
8. http://www.rollsroyce.com/marine/products/electrical_power_systems/storage/index.jsp

9. Dûrr M (2006) Dynamic model of a lead acid battery for use in a domestic fuel cell system. J Power Sources 161:1400–1411
10. Benyahia N, Benamrouche N, Rekioua T (2012) Modeling, design and simulation of fuel cell modules for small marine applications. 20th International Conference on Electrical Machines, ICEM 2012:1–6
11. Zoroofi S (2008) Modeling and simulation of vehicular power systems. Thesis of master, Chalmers University of Technology, Gothenburg
12. Marquet A (2003) Storage of electricity in electric systems. Eng Tech 4:030
13. Francois CJ (1997/1998) Modeling the art of charge electric vehicle batteries. Faculty of Applied Sciences Degree Legal Electricians Civil Engineer, University of Liege, Liege
14. Multon B, Ben Ahmed H, Bernar N, Kerzreho C (2007) The inertial storage electromechanics. 3EI J 48:18–29
15. Salameh ZM, Casacca MA, Lynch WA (1992) A mathematical model for lead-acid batteries. IEEE Trans Energy Convers 7(1):93–98
16. EA Technology (2004) Review of electrical energy storage technologies and systems and of their potential for the UK. www.wearemichigan.com/JobsAndEnergy/documents/file15185.pdf, pp 1–34
17. Kosin L, Usach F (1995) Electric characteristics of lead battery. Russ J Appl Chem 143(3):1–4
18. Labbé J (2006) Electrolytic hydrogen as a storage of electricity for photovoltaic systems insulated. Ph.D. thesis, Specialty Energetic, Paris School of Mines, Paris
19. Francisco M, Longatt G (2006) Circuit based battery models: a review. In: 2do Congreso Iberoamericano De Estudiantes De Ingeniería Eléctrica (II CIBELEC 2006), pp 1–5
20. Chan HM, Stutanto D (2000) A new battery model for use with battery energy storage systems and electric vehicles power system. IEEE Power Eng Soc Winter Meet 1:470–475
21. Ceraolo M (2000) Dynamical models of lead-acid batteries. IEEE Trans Power Syst 15:1184–1190
22. http://www.windpowerengineering.com/design/electrical/power-storage/solving-the-use-it-or-lose-it-wind-energy-problem/
23. http://www.windpowerengineering.com/design/electrical/power-storage/overview-of-wind-power-storage-media/
24. Ibrahim H, Ghandour M, Dimitrova M, Ilinca A, Perron J (2011) Integration of wind energy into electricity systems: technical challenges and actual solutions. Energy Procedia 6:815–824
25. Seyoum D, Rahman MF, Grantham C (2003) Terminal voltage control of a wind turbine driven isolated induction generator using stator oriented field control. In: Eighteenth annual IEEE applied power electronics conference and exposition (APEC'03), vol 2. Miami Beach, pp 846–852, 9–13 Fevrier 2003
26. Hansen AD, Sørensen P, Iov F, Blaabjerg F (2006) Centralised power control of wind farm with doubly fed induction generators. Renew Energy 31(7):935–951
27. Jiao L, Ooi B-T, Joós G, Zhou F (2005) Doubly-fed induction generator (DFIG) as a hybrid of asynchronous and synchronous machines. Electr Power Syst Res 76(1–3):33–37
28. Katiniotis IM, Ioannides MG, Vernados PG (2005) Operation of induction generator in the magnetic saturation region as a self-excited and as a double output system. J Mate Process Technol 161(1):263–268
29. Singh GK (2004) Self-excited induction generator research—a survey. Electr Power Syst Res 69(2–3):107–114
30. Tapia A, Tapia G, Ostolaza JX (2004) Reactive power control of wind farms for voltage control applications. Renew Energy 29(3):377–392
31. Valtchev V, Van den Bossche A, Ghijselen J, Melkebeek J (2000) Autonomous renewable energy conversion system. Renew Energy 19(1):259–275

32. Bhatti TS, Al-Ademi AAF, Bansal NK (1997) Load frequency control of isolated wind diesel hybrid power systems. Energy Convers Manag 38(9):829–837
33. Djurovic M, Joksimovic G (1996) Optimal performance of double fed induction generator in windmills. Renew Energy 9(1–4):862–865
34. Idjdarene K, Rekioua D, Rekioua T, Tounzi A (2011) Wind energy conversion system associated to a flywheel energy storage system. Analog Integr Circ Sig Process 69(1):67–73

Chapter 5
Control of Wind Turbine Systems

5.1 Basic Principles of Wind Turbine Control Systems

The time constants of the electrical system are much lower than that of other parts of the aerogenerator. This allows to separate the control of the electrical machine from the aeroturbine and to define two levels of control.

- A control module of level 1 is applied to the electric generator via the power converter and the pitch system.
- A control module of level 2 which provides inputs β^* and T^*_{em} of level 1. This control configuration is shown in Fig. 5.1.

Control level 2 of the aerogenerator can be performed without modeling the electrical part. This approach is often used in the specific literature to the aerogenerator control. However, we can find many works dedicated specifically to the electrical part control we can mention.

5.2 Level 1 (Mechanical Part)

5.2.1 No Linear Control by Static State Feedback

We assume that the wind speed is measured. The control laws developed here can be applied to large wind turbines. $\tilde{v}_{\mathrm{wind}}$ is a fictive measure of the wind speed. We can rebuild the aerodynamic torque from this measure and that of the rotor speed $\tilde{\omega}_t$

$$\tilde{T}_a = \frac{1}{2}\rho\pi R^3 C_q(\tilde{\lambda})\tilde{v}^2_{\mathrm{wind}} \tag{5.1}$$

D. Rekioua, *Wind Power Electric Systems*, Green Energy and Technology,
DOI: 10.1007/978-1-4471-6425-8_5, © Springer-Verlag London 2014

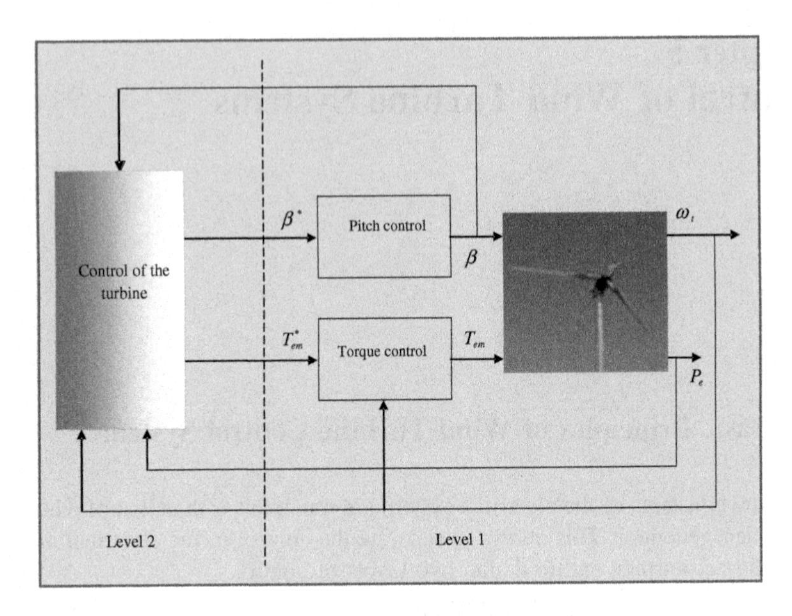

Fig. 5.1 Control levels of a variable wind speed

with

$$\tilde{\lambda} = \frac{\tilde{\omega}_t R}{\tilde{v}_{\text{wind}}} \tag{5.2}$$

It is also assumed that the rotor speed ω_t measurement is available. According to the dynamic equations of the system, we have:

$$\dot{\omega}_t = \frac{1}{J_t} T_a - \frac{K_t}{J_t} \omega_t - \frac{1}{J_t} T_g \tag{5.3}$$

We will use a linearizing control T_g which reduces the system to a single integrator with a new input w

$$\dot{\omega}_t = w \tag{5.4}$$

That control is given by:

$$T_g = J_t \left[\frac{1}{J_t} T_a - \frac{K_t}{J_t} \omega_t - w \right] \tag{5.5}$$

Tracking error is defined by:

$$\varepsilon_\omega = \omega_{t_{\text{opt}}} - \omega_t \tag{5.6}$$

where

$$\omega_{t_{\rm opt}} = \frac{\lambda_{\rm opt} v_{\rm wind}}{R} \tag{5.7}$$

$\omega_{t_{\rm opt}}$ is the rotor's speed which allows to have a specific speed corresponding to the optimal value of the power coefficient $C_p(\lambda)$.

We want to impose to the error ε_ω a dynamic of first order

$$\dot{\varepsilon}_\omega + a_0\,\varepsilon_\omega = 0, \quad a_0 > 0 \tag{5.8}$$

Similarly, we obtain the expression

$$w = \dot{\omega}_{t_{\rm opt}} + a_0\left(\omega_{t_{\rm opt}} - \omega_t\right) \tag{5.9}$$

Hence, the control:

$$T_g = T_a - K_t\omega_t - a_0 J_t \varepsilon_\omega - J_t \dot{\omega}_{t_{\rm opt}} \tag{5.10}$$

The choice of the dynamic of the tracking error of the first order is due to the fact that the relative degree of the system is equal to 1.

This control by static state feedback leads to good feedback in the lack of disturbance, but if not, it has the disadvantage of not rejecting these disturbances.

We make an application under MATLAB/SIMULINK. The simulation was performed under the following conditions:

- Lack of additive disturbance to the generator torque.
- A profile of an average wind velocity of 7 m/s.

The characteristics of the controlled wind turbine are shown in Fig. 5.1, which are the wind speed, the aerodynamic torque, rotor speed and its optimal reference, the aerodynamic power and its optimal reference, and the torque/couple of generator. We note that the rotor speed is very close to its optimal value. The wind energy capture is maximal in the range of 0–150 s. The trajectory of the aerodynamic power is almost identical to the one of its reference; then, we can confirm that this control method gives good feedback in the lack of disturbances.

5.2.2 No Linear Dynamic Control by State Feedback

This control is developed with the objective of rejecting the constant additive disturbances on the control. We apply a second-order dynamic on the tracking error formula: $\varepsilon_\omega = \omega_{t_{\rm opt}} - \omega_t$

$$\ddot{\varepsilon}_\omega + b_1\dot{\varepsilon}_\omega + b_0\,\varepsilon_\omega = 0 \tag{5.11}$$

b_0 and b_1 are selected such as the polynomial $s^2 + b_1 s + b_0 = 0$ is Hurwitz.

Assuming also a constant disturbance acts on the system, we have:

$$J_t\dot{\omega}_t = T_a - K_t\omega_t - T_g + d \tag{5.12}$$

Deriving this equation, we obtain:

$$\ddot{\omega}_t = \frac{1}{J_t}\left[\dot{T}_a - K_t\dot{\omega}_t - \dot{T}_g\right] \tag{5.13}$$

We are looking for a control that will change the system to a double integrator with the new input w

$$\ddot{\omega}_t = w \tag{5.14}$$

Then, we deduce

$$\dot{T}_g = J_t\left[\frac{1}{J_t}\dot{T}_a - \frac{K_t}{J_t}\dot{\omega}_t - w\right] \tag{5.15}$$

We use Eqs. (5.11) and (5.13) and obtain :

$$w = \ddot{\omega}_{t_{\text{opt}}} + b_1\left(\dot{\omega}_{t_{\text{opt}}} - \dot{\omega}_t\right) + b_0\left(\omega_{t_{\text{opt}}} - \omega_t\right) \tag{5.16}$$

Finally, we obtain the dynamic control:

$$\dot{T}_g = \dot{T}_a + (b_1J_t - K_t)\dot{\omega}_t + b_0J_t\omega_t - J_t\left(\ddot{\omega}_{t_{\text{opt}}} + b_1\dot{\omega}_{t_{\text{opt}}} + b_0\omega_{t_{\text{opt}}}\right) \tag{5.17}$$

The compromise between the optimization of wind energy capture and the minimization of the transitional efforts undergone by the driving device is accomplished by choosing a tracking dynamic which permits to follow of the average trend of the optimal rotation speed without following closely the peaks of wind (Fig. 5.2).

We make an application under MATLAB/SIMULINK with the same conditions in no linear control by static state feedback (NLCSSF). The controlled wind turbine characteristics are shown in Fig. 5.3. We note that the rotor speed increases with the wind speed. The wind speed trajectory is almost identical to the one of its reference. The captured aerodynamic power is maximal for the wind high speeds, and it is noted too that it coincides with the optimal power.

5.2.3 *Indirect Speed Control*

The wind system, under certain conditions, is dynamically stable around any balance point of the maximal efficiency curve for a generator torque and a constant operating wind speed. The maximal aerodynamic efficiency curve is defined in the

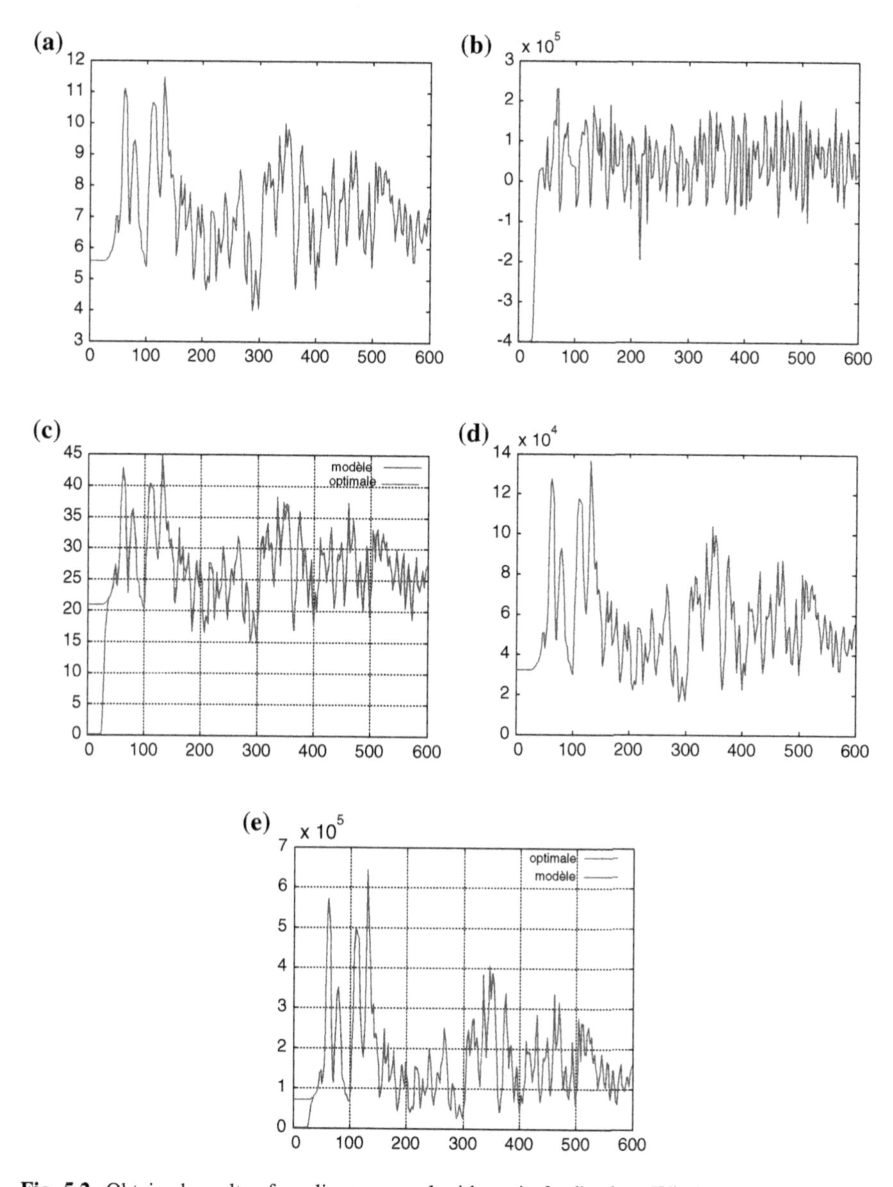

Fig. 5.2 Obtained results of nonlinear control with static feedback. **a** Wind speed profile v_{wind}. **b** Generator torque T_g. **c** Rotor speed ω_t. **d** Aerodynamic torque T_a. **e** Aerodynamic power P_a

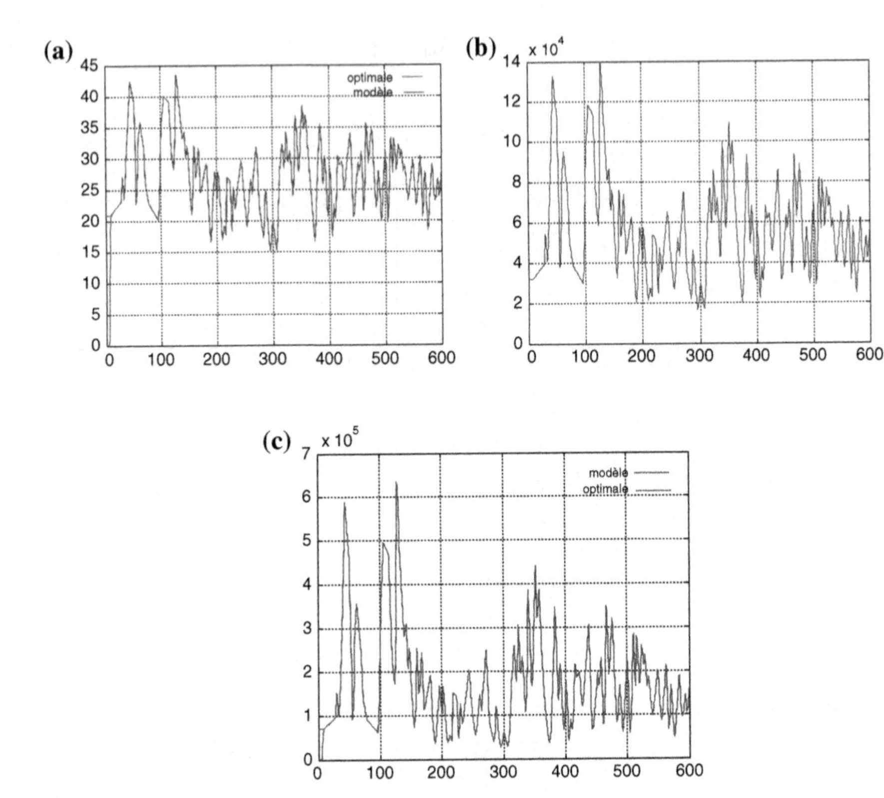

Fig. 5.3 Obtained results of the nonlinear control by dynamic state feedback. **a** Rotor speed ω_t. **b** Aerodynamic torque T_a. **c** Aerodynamic power P_a

plane (ω_t, T_a) by the set of points $\left(\omega_{t_{\text{opt}}}, T_{a_{\text{opt}}}\right)$ corresponding to the interval of wind speeds in which the wind turbine operates.

We have:

$$T_a = \frac{1}{2}\rho\pi R^3 C_q(\lambda_{\text{opt}})v_{\text{wind}}^2 \tag{5.18}$$

If the aerodynamic torque is controlled so as to follow the optimal torque, the wind turbine remains around its optimal yield curve.

We have a given point $\left(\lambda_0, C_{p_0}\right)$ of the curve $C_p(\lambda)$ that we want to follow. In order to maximize the energy production below the nominal power P_{nom}, this point is selected in a neighborhood where power coefficient is at its maximum $\left(\lambda_{\text{opt}}, C_{p_{\text{opt0}}}\right)$.

Since

$$C_q(\lambda) = \frac{C_p(\lambda)}{\lambda} \tag{5.19}$$

The aerodynamic torque could be written:

$$T_a = \frac{1}{2}\rho\pi R^3 \frac{C_p(\lambda)}{\lambda} v_{\text{wind}}^2 \tag{5.20}$$

If $\lambda = \lambda_{\text{opt}}$, T_a could be expressed according to the rotor

$$T_a = \frac{1}{2}\rho\pi R^5 C_{p_{\text{opt}}} \frac{1}{\lambda_{\text{opt}}^3} \omega_t^2 \tag{5.21}$$

This torque corresponds to an optimal operating in relation to the wind speed, and it is proportional to the square of the aeroturbine rotation speed at the operating point $\left(\lambda_{\text{opt}}, C_{p_{\text{opt}0}}\right)$.

$$T_a = k_{\text{opt}}\omega_t^2 \tag{5.22}$$

with

$$k_{\text{opt}} = \frac{\rho}{2}\pi R^5 C_{p_{\text{opt}}} \frac{1}{\lambda_{\text{opt}}^3} \tag{5.23}$$

If we consider the model with one mass of wind turbine, in steady state we have:

$$0 = T_a - K_t\omega_t - T_g \tag{5.24}$$

The generator torque T_g satisfies:

$$T_g = k_{\text{opt}}\omega_t^2 - K_t\omega_t \tag{5.25}$$

The structure of the indirect control speed is shown in Fig. 5.4.

This technique has two disadvantages: The first is that it does not consider sufficiently the dynamic aspects of the aeroturbine and the wind. Indeed, the synthesis of this control assumes that the wind turbine is in a steady state on the optimal efficiency curve. The rotor speed fluctuations in response to the wind variations deviate in a significant way the wind turbine from this trajectory. In addition, the wind speed variations are faster than the closed-loop dynamic system; the control system does not have the time to stabilize on the optimal efficiency curve. This continuous transaction is accompanied by energy losses. The second disadvantage is the lack of robustness toward the measurement noise and disturbances. We make an application under MATLAB/SIMULINK. The simulation is performed under the

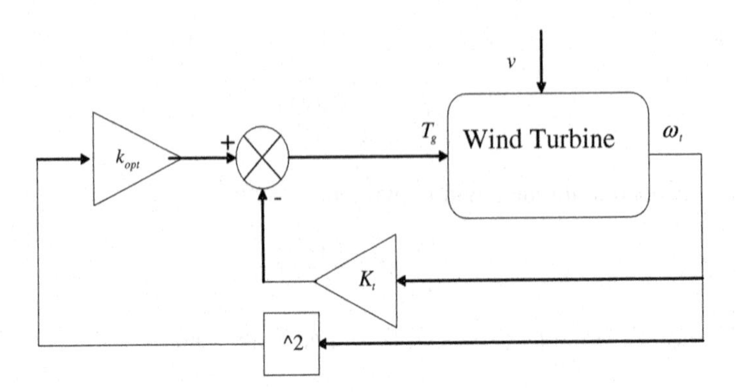

Fig. 5.4 Indirect control speed

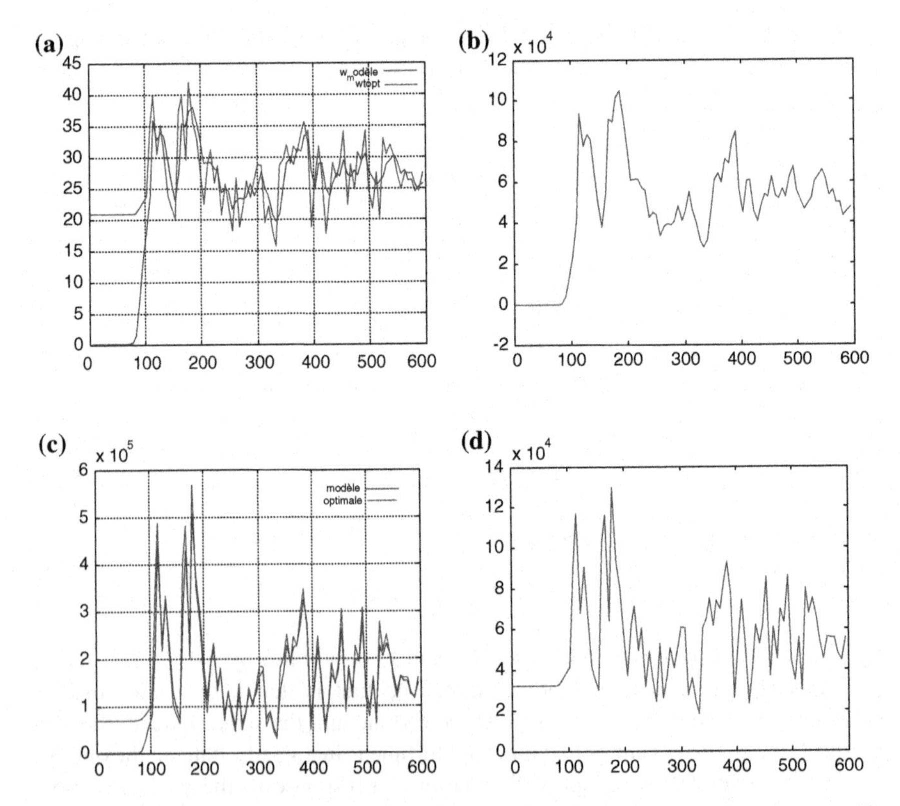

Fig. 5.5 Obtained results by indirect control speed. **a** Rotor speed ω_t. **b** Aerodynamic torque T_a. **c** Aerodynamic power P_a. **d** Aerodynamic torque T_a

same conditions as for the previous two methods. We observe from the Fig. 5.5a that the rotor speed deviates from its optimal trajectory even in the lack of disturbance. The aerodynamic power (Fig. 5.5d) is consequently affected by this deviation.

5.2.4 Comparison Between the Three Controls

To highlight the most effective controls, we make a comparison between the three methods under two different perturbation torques.

- For Tr = 10 kN m

	ISV	NLCSSF	NLDCSF
Static error on P_a (kW)	14.7	2.9	0
Static error on ω_t (tr/min)	2.97	0.47	0

- For Tr = 15 kN m

	ISV	NLCSSF	NLDCSF
Static error on P_a (kW)	22.2	4.15	0
Static error on ω_t (tr/min)	4.22	0.72	0

We can remark in the studied example that the most effective strategy is the nonlinear dynamic control by state feedback (NLDCSF). It rejects the disturbance on the control and achieves better tracking of the optimal speed.

5.3 Level 2 (Electrical Part)

5.3.1 Scalar Control of Wind System

The steady-state performance of an induction motor is modeled using the conventional equivalent circuit (Fig. 5.6). Assuming that R_m is infinite, the expression of electromagnetic can be written as:

$$T_{\mathrm{emAC}} = 3P \cdot \left(\frac{V_s}{\omega_s}\right)^2 \frac{\frac{R_r}{g}\omega_s X_m^2}{\left[\frac{R_r R_s}{g} - \omega_s^2 \cdot \left(L_s L_r - X_m^2\right)\right]^2 + \omega_s^2\left[\frac{R_r L_s}{g} + R_s L_r\right]^2} \tag{5.26}$$

with

$$g = \frac{\omega_s - \omega}{\omega_s} \tag{5.27}$$

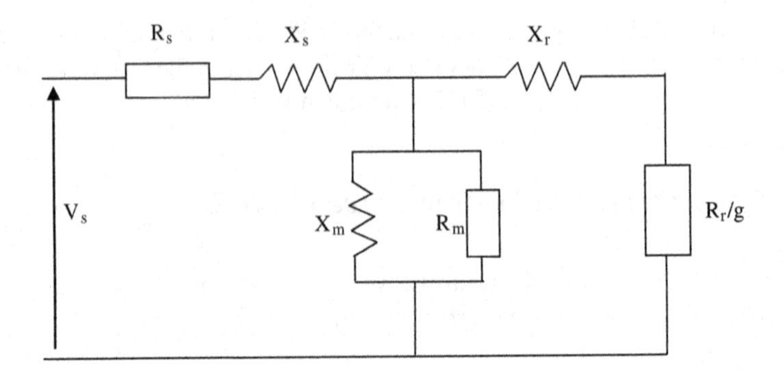

Fig. 5.6 Induction motor equivalent circuit

V_s rms motor voltage (V). R_s stator resistance per phase (Ω), R_r equivalent rotor resistance per phase (Ω), R_m core loss resistance (Ω), X_s stator leakage reactance (Ω), X_r equivalent rotor leakage reactance (Ω), X_m magnetizing reactance (Ω), ω_s angular frequency of the supply (rd/s), ω motor speed (rd/s), and g slip.

Mechanical power P_{mec}, stator current I_s, and stator power P_s will be:

$$P_{\text{mec}} = 3\frac{R_r}{g}I_r^2 \tag{5.28}$$

$$I_s = \left|1 + \frac{R_r/g + jL_r\omega_s}{jL_m\omega_s}\right| I_r \tag{5.29}$$

$$P_s = P_{\text{mec}} + P_{jr} + P_{js} \tag{5.30}$$

From the three previous equations, we obtain the total power absorbed by the stator according to speed:

$$P_s = 3k^2\left(\frac{(\omega_s - \omega)\cdot\omega_s}{R_r\left(\frac{L_s}{X_m}\right)^2} + R_s\left(\frac{R_r^2 + L_r^2\cdot(\omega_s - \omega)^2}{(R_rL_s)^2}\right)\right) \tag{5.31}$$

Generally, in variable speed drives, motor airgap flux is maintained constant at all frequencies so that the motor can deliver a constant torque. This will occur if the V_s/f_s (V_s/ω_s) ratio is kept constant at its nominal value. To compensate the voltage drop due to stator resistance effect at low frequencies, a boost voltage V_{s0} is added to phase voltages. For aerodynamic loads, the stator voltage as function of frequency is given by (Fig. 5.7):

$$\text{For}\quad 0 \le f \le f_N \quad V_s = V_{s0} + K\cdot f_N \tag{5.32}$$

$$\text{For}\quad f \ge f_N \quad V_s = V_N \tag{5.33}$$

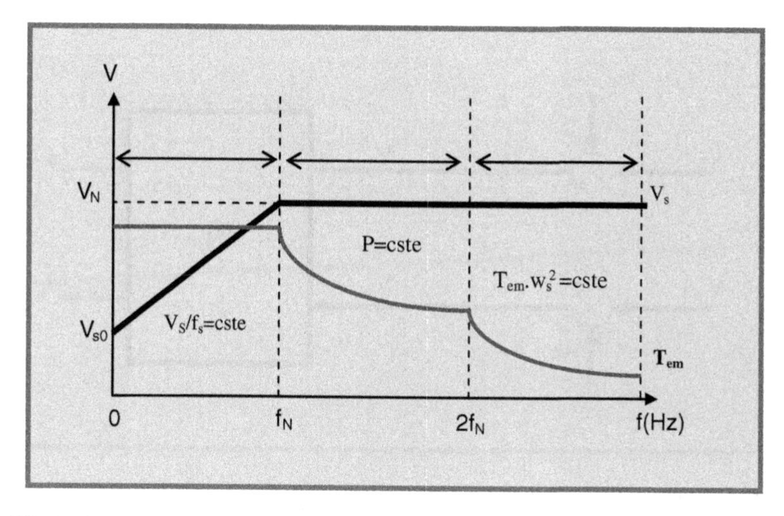

Fig. 5.7 Scalar control [7]

5.3.2 Vector Control of Wind System

We oriented the rotor flux Φ_r along the direct axis

$$\begin{cases} \Phi_{dr} = \Phi_r \\ \Phi_{qr} = 0 \end{cases} \tag{5.34}$$

We obtain:

$$\begin{cases} V_{ds} = R_s I_{ds} + \sigma L_s \dfrac{\mathrm{d}I_{ds}}{\mathrm{d}t} + \dfrac{L_m}{L_r}\dfrac{\mathrm{d}\Phi_r}{\mathrm{d}t} - \omega_s \cdot \sigma L_s \cdot I_{qs} \\ V_{qs} = R_s I_{qs} + \sigma L_s \dfrac{\mathrm{d}I_{qs}}{\mathrm{d}t} + \omega_s \dfrac{L_m}{L_r}\Phi_r + \omega_s \cdot \sigma L_s \cdot I_{ds} \\ \omega r\mathrm{AC} = \dfrac{L_m}{T_r}\dfrac{I_{qs}}{\Phi_r} \\ T_{\mathrm{emAC}} = p\dfrac{L_m}{L_r}(\Phi_r \cdot I_{qs}) \end{cases} \tag{5.35}$$

The new control is as follows (Fig. 5.8):

$$\begin{cases} V_{ds}{}^{*} = (R_s + s \cdot \sigma L_s) I_{ds} = V_{ds} + \omega_s \cdot \sigma L_s \cdot I_{qs} = V_{ds} + e_{ds} \\ V_{qs}{}^{*} = (R_s + s \cdot \sigma L_s) I_{qs} = V_{qs} - \left(\omega_s \dfrac{L_m}{L_r}\Phi_r + \omega_s \cdot \sigma L_s \cdot I_{ds} \right) = V_{qs} - e_{qs} \end{cases} \tag{5.36}$$

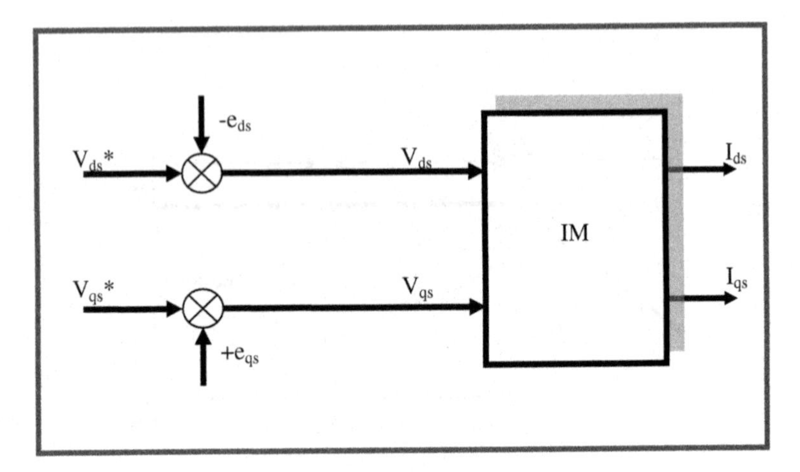

Fig. 5.8 Structure of the decoupling regulator

5.3.3 Direct Torque Control of Wind System

5.3.3.1 DTC Principals

Direct torque control (DTC) of induction machines (IM) is a powerful control method for motor drives. Featuring a direct control of the stator flux and torque instead of the conventional current control technique, it provides a systematic solution to improve operating characteristics of the motor and the voltage inverter source [1–10]. In principle, DTC method is based mainly on instantaneous space vector theory. By optimal selection of the space voltage vectors in each sampling period, DTC achieves effective control of the stator flux and torque. Consequently, the number of space voltage vectors and switching frequency directly influence the performance of DTC control system. For a prefixed switching strategy, the drive operation, in terms of torque, switching frequency, and torque response are quite different at low and high speed.

5.3.3.2 DTC Structure

A configuration of DTC scheme is represented in Fig. 5.9. In this system, the instantaneous values of flux and torque can be calculated from stator variables and mechanical speed or using only stator variables. Stator flux and torque can be controlled directly and independently by properly selecting the inverter switching configurations. With a three-phase voltage source inverter, six nonzero-voltage vectors and two zero-voltage vectors can be applied to the machine terminals.

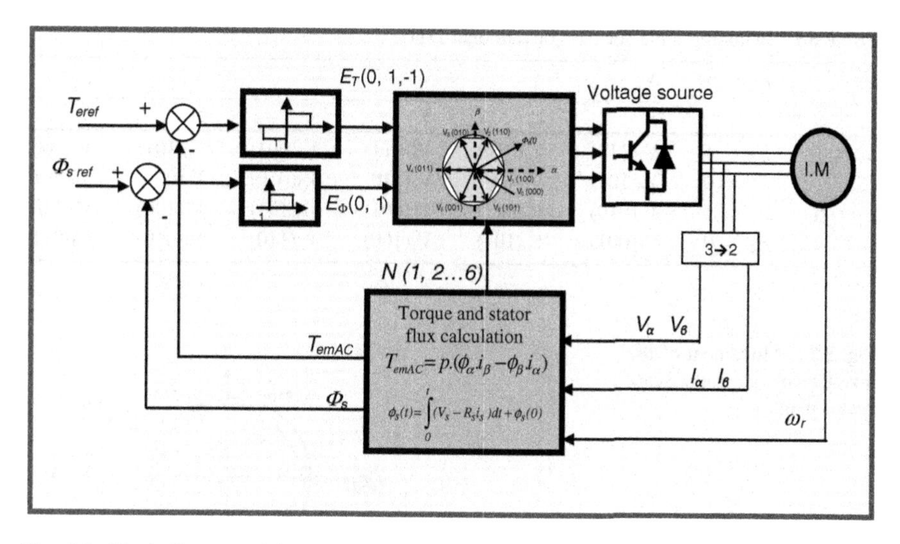

Fig. 5.9 Block diagram of the conventional DTC of induction motor drives

The stator flux can be estimated using measured current and voltage vectors [11]:

$$\phi_s(t) = \int_0^t (V_s - R_s i_s)\mathrm{d}t \tag{5.37}$$

Since stator resistance R_s i_s is relatively small, the voltage drop R_s i_s might be neglected ($V_s \gg R_s i_s$); we obtain:

$$\phi_s(t) = V_s \cdot T + \phi_s(0) \tag{5.38}$$

$\phi_s(0)$ is the stator flux initial value at the switching time and T the sampling period in which the voltage vector is applied to stator windings.

It is clear that stator flux directly depended on the space voltage vector V_s and the system sampling period T.

The stator voltage vector V_s is selected using Table 5.1, where signs of torque and flux errors E_T and E_Φ are determined with a zero hysteresis band (Fig. 5.9).

$$E_T = T_{\text{eref}} - T_{\text{emAC}} \tag{5.39}$$

$$E_\phi = \phi_{\text{sref}} - \phi_s \tag{5.40}$$

where

$$\phi_s = \sqrt{(\phi_{s\alpha})^2 + (\phi_{s\beta})^2} \tag{5.41}$$

Table 5.1 Switching table for the conventional DTC

E_T	E_ϕ	N					
		1	2	3	4	5	6
$E_T = 1$	$E_\phi = 1$	$V_2(110)$	$V_3(010)$	$V_4(011)$	$V_5(001)$	$V_6(101)$	$V_1(100)$
	$E_\phi = 0$	$V_6(101)$	$V_1(100)$	$V_2(110)$	$V_3(010)$	$V_4(011)$	$V_5(001)$
$E_T = 0$	$E_\phi = 1$	$V_3(010)$	$V_4(011)$	$V_5(001)$	$V_6(101)$	$V_1(100)$	$V_2(110)$
	$E_\phi = 0$	$V_5(001)$	$V_6(101)$	$V_1(100)$	$V_2(110)$	$V_3(010)$	$V_4(011)$

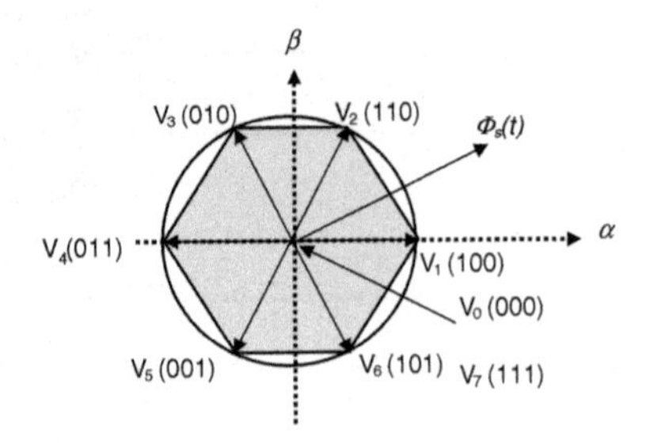

Fig. 5.10 Movement of the inverter voltage in the space vector plane

Table 5.1 shows the associated inverter switching states of the conventional direct torque control strategy.

The definition of flux sector and inverter voltage vectors is shown in Fig. 5.10, where the stator flux vector is rotating with a speed of ω_{rAC}. For each possible switching configuration, the output voltages can be represented in terms of space vector, according to the following equation:

$$V_S = V_{S\alpha} + jV_{S\beta} = \sqrt{\frac{2}{3}}\left[V_a + V_b \exp\left(j\frac{2\pi}{3}\right) + V_c \exp\left(j\frac{4\pi}{3}\right)\right] \tag{5.42}$$

where V_a, V_b, and V_c are voltage phases

5.3.3.3 Application: DTC of IM Fed by a Wind Turbine

The DTC control applied to the induction motor is simulated through MATLAB/SIMULINK. The simulation model of the induction motor is given in Fig. 5.11.

Simulation diagram of the stator flow control stator is illustrated in Fig. 5.12. The control signal flow (fx_con) is generated by a comparator with hysteresis,

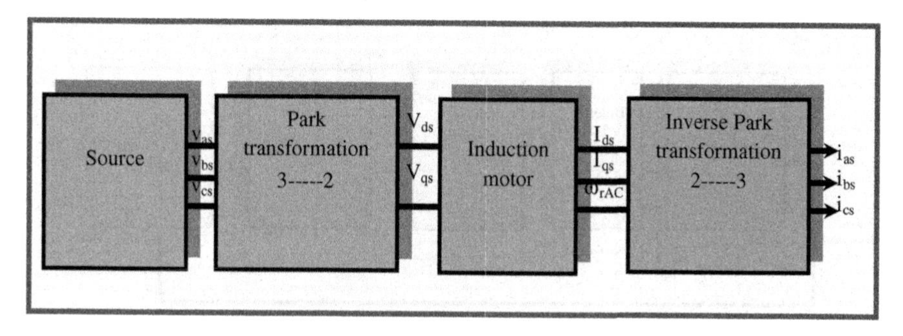

Fig. 5.11 Simulation diagram of induction motor

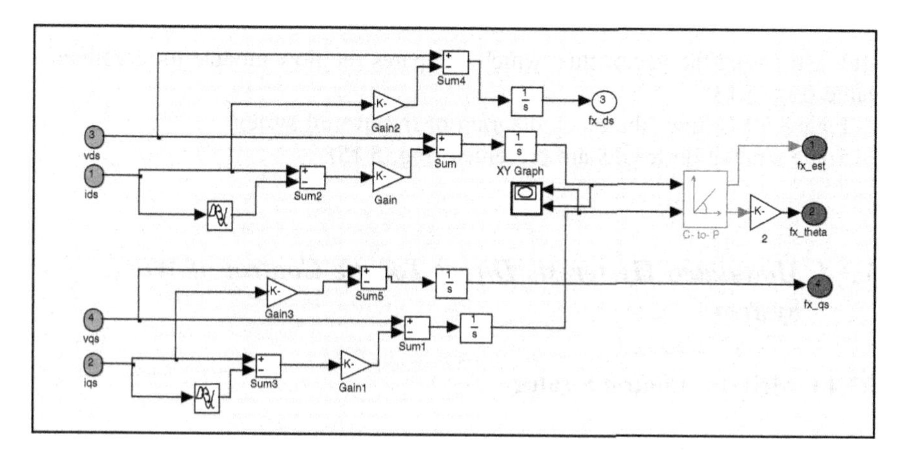

Fig. 5.12 Simulation diagram of stator flux control

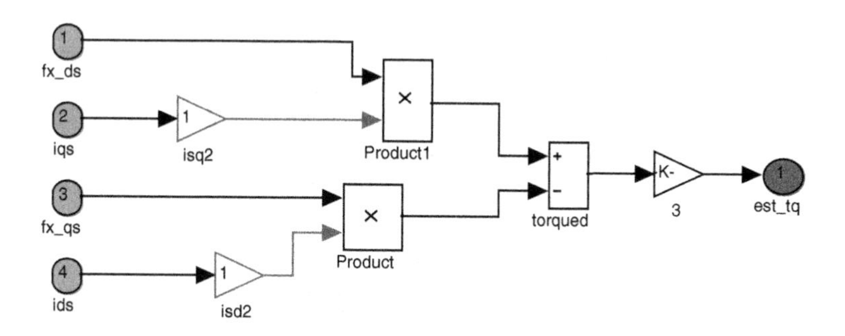

Fig. 5.13 Simulation diagram of the electromagnetic torque control

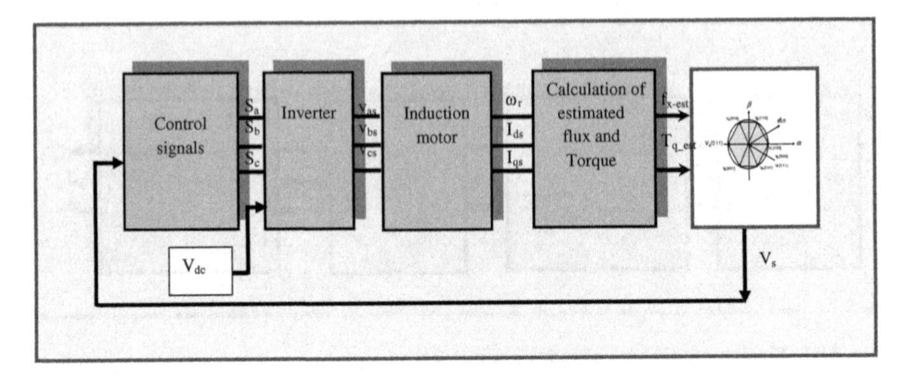

Fig. 5.14 Simulation diagram of the overall structure of the DTC control

after estimating the vector flux, which compares the flow module to its reference value (Fig. 5.13).

Figure 5.14 shows the block diagram of the overall system.

Some simulation results are presented (Fig. 5.15).

5.3.4 Modulated Hysteresis Direct Torque Control of Wind System

5.3.4.1 MHDTC Control Strategy

The use of hysteresis controllers in DTC leads to distortion in the currents supplied to the grid and large variations in the power factor. In another hand, when using vector control strategy, the direct bus voltage varies with the wind variations, but the associated PWM leads to a better quality of current and power factor. So, in order to obtain performances with the advantages of both DTC and FOC methods, we apply an improved modulated hysteresis direct torque control which consists in superposing, to the constant torque reference, a triangular signal with the desired switching frequency as in the PWM control used with FOC. The modulated reference torque is compared to the estimated torque by using a hysteresis controller as in the classical DTC (Fig. 5.16). The new torque reference T^*_{eref} is then defined by:

$$T^*_{\text{eref}} = T_{\text{eref}} + T_{\text{etr}} \tag{5.43}$$

where T_{eref} is the reference torque and T_{etr} the triangular signal.

To impose a switching frequency, the estimated torque variation during a half period should not exceed the difference between the maximum of the upper limit and the minimum of the lower limit. In this case, the switching frequency is equal

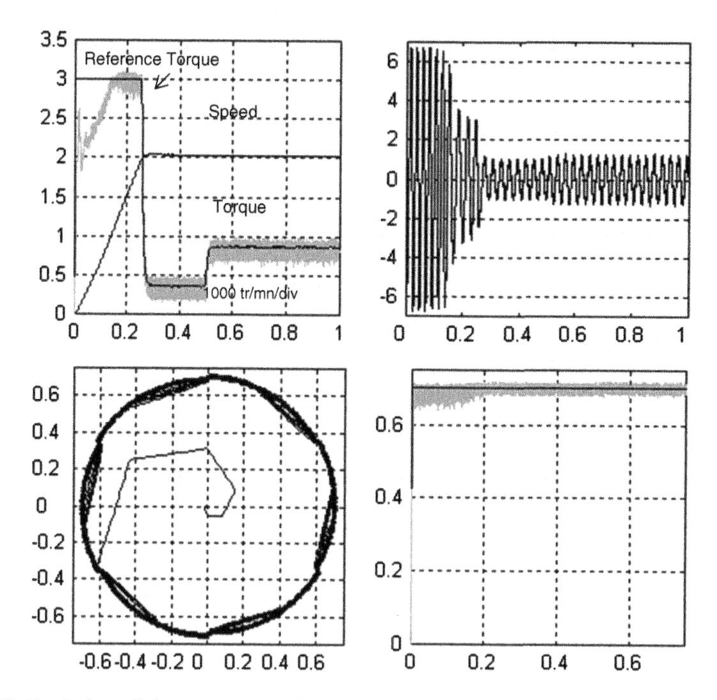

Fig. 5.15 Evolution of the torque, speed, stator current, and stator flux with a speed reference

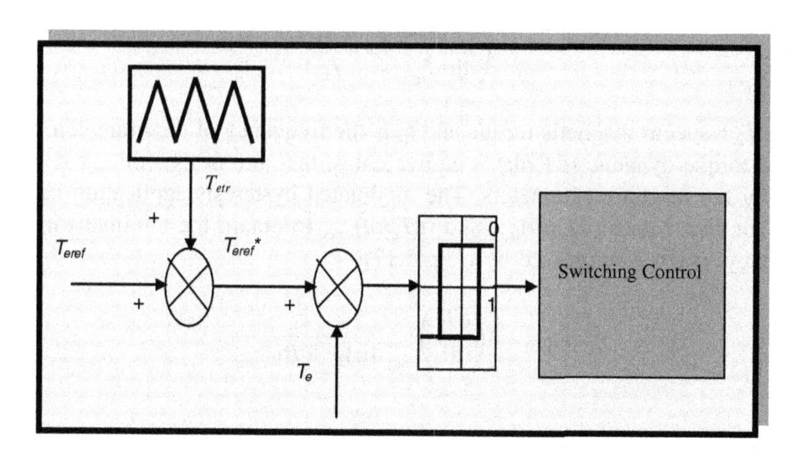

Fig. 5.16 Basic concept of the modulated hysteresis

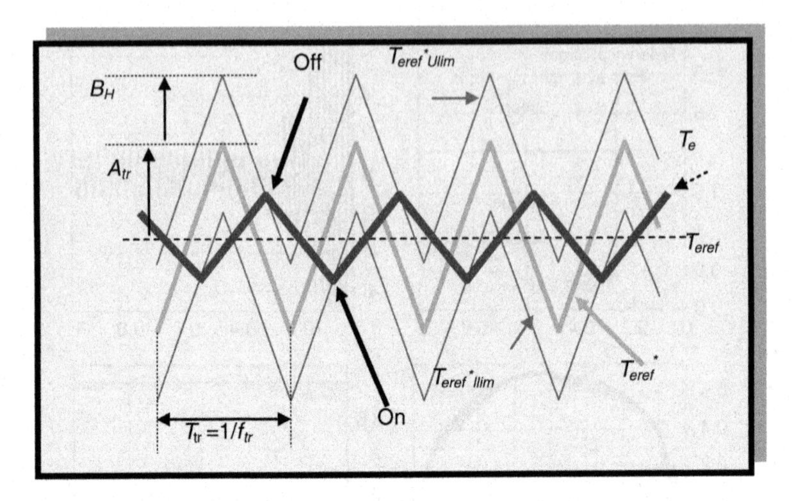

Fig. 5.17 Determination of the switch states [5]

to the imposed one only if the triangular signal magnitude A_{tr} and the hysteresis bandwidth B_H verify the two conditions:

$$\left(\frac{\mathrm{d}T_e}{\mathrm{d}t}\right)_{\max} = 4f_{tr}(A_{tr} + B_H) \tag{5.44}$$

$$\left(\frac{\mathrm{d}T_e}{\mathrm{d}t}\right)_{\min} > \frac{2B_H}{T_{tr}} \tag{5.45}$$

where T_e is electromagnetic torque and f_{tr} is the frequency of the triangular signal.

The torque dynamic ($\mathrm{d}T_e/\mathrm{d}t$) is equivalent to the current dynamic, and so it is fixed by the machine parameters. The modulated hysteresis application needs to know the two values $(\mathrm{d}T_e/\mathrm{d}t)_{\min}$ and $(\mathrm{d}T_e/\mathrm{d}t)_{\max}$ to obtain the minimal triangular frequency which is expressed by (Fig. 5.17):

$$f_{tr\min} = \left(\frac{\mathrm{d}T_e}{\mathrm{d}t}\right)_{\max} \frac{1}{4(A_{tr} + B_H)_{imp}} \tag{5.46}$$

5.3.4.2 Application to Wind Energy Conversion System

We make an application of MHDTC on WECS. The studied system is represented in the Fig. 5.18.

The control principle consists in adjusting the active power supplied to the grid to its reference value P_{ref} and the reactive power Q_{ref} to zero in order to fix the power factor at the unit. The active power reference is deduced by controlling the

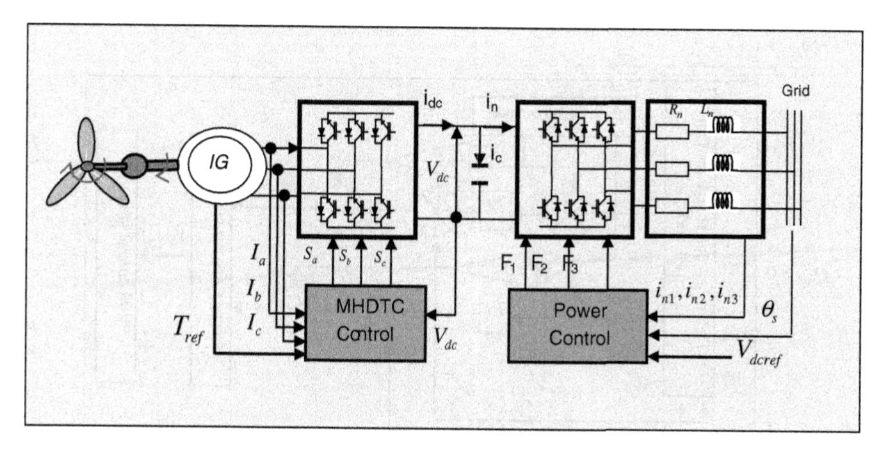

Fig. 5.18 Wind generator based on an induction generator

direct bus voltage with a proportional integral corrector generating the current reference $i_{c-\mathrm{ref}}$ to the capacitance.

Thus, we can write:

$$P_{\mathrm{ref}} = V_{\mathrm{dc}} \cdot (i_{\mathrm{dc}} - i_{\mathrm{cref}}) \tag{5.47}$$

$$P_{\mathrm{ref}} = P_{\mathrm{dc}} - P_{\mathrm{cref}} \tag{5.48}$$

with

$$i_{\mathrm{cref}} = PI \cdot (V_{\mathrm{dcref}} - V_{\mathrm{dc}}) \tag{5.49}$$

Active and reactive power references P_{ref} and Q_{ref} are given by the following equations:

$$P_{\mathrm{ref}} = E_d i_{\mathrm{nd\_ref}} + E_q i_{\mathrm{nq\_ref}} \tag{5.50}$$

$$Q_{\mathrm{ref}} = E_q i_{\mathrm{nd\_ref}} - E_d i_{\mathrm{nq\_ref}} \tag{5.51}$$

where E_d, E_q are the Park transform of the stator emfs E_1, E_2, E_3

Equation (5.50) is then multiplied by E_d and Eq. 5 by E_q. Then, the addition and subtraction of the two new equations give the reference current values according to active and reactive power ones by:

$$i_{\mathrm{nd-ref}} = \frac{P_{\mathrm{ref}} E_{\mathrm{d}} + Q_{\mathrm{ref}} E_q}{E_d^2 + E_q^2} \tag{5.52}$$

$$i_{\mathrm{nq-ref}} = \frac{P_{\mathrm{ref}} E_q - Q_{\mathrm{ref}} E_d}{V_d^2 + V_q^2} \tag{5.53}$$

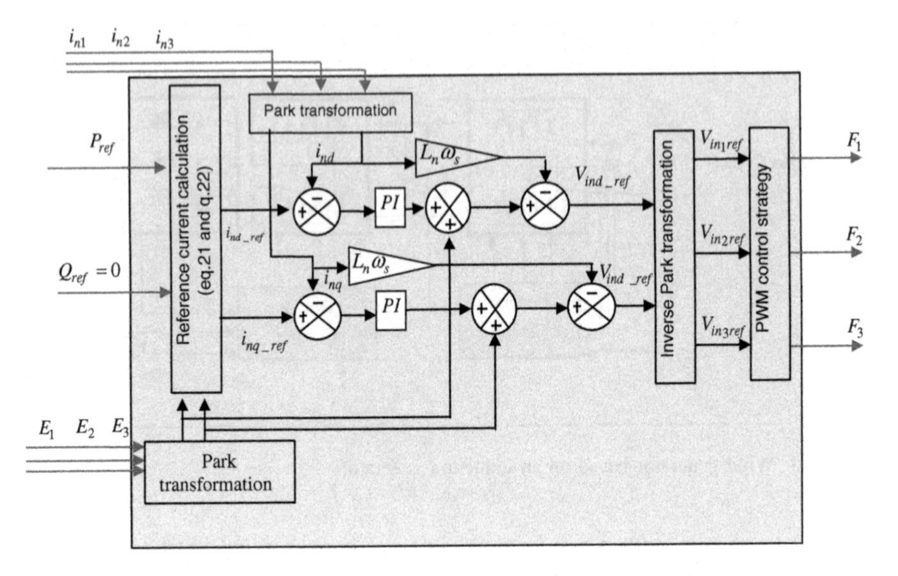

Fig. 5.19 Power control [5]

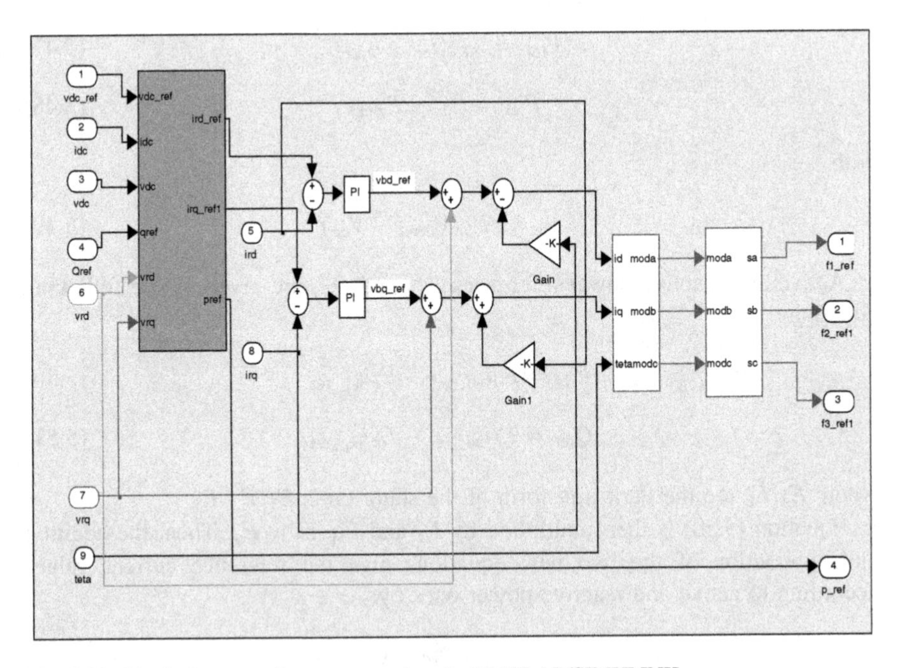

Fig. 5.20 Block diagram of power control under MATLAB/SIMULINK

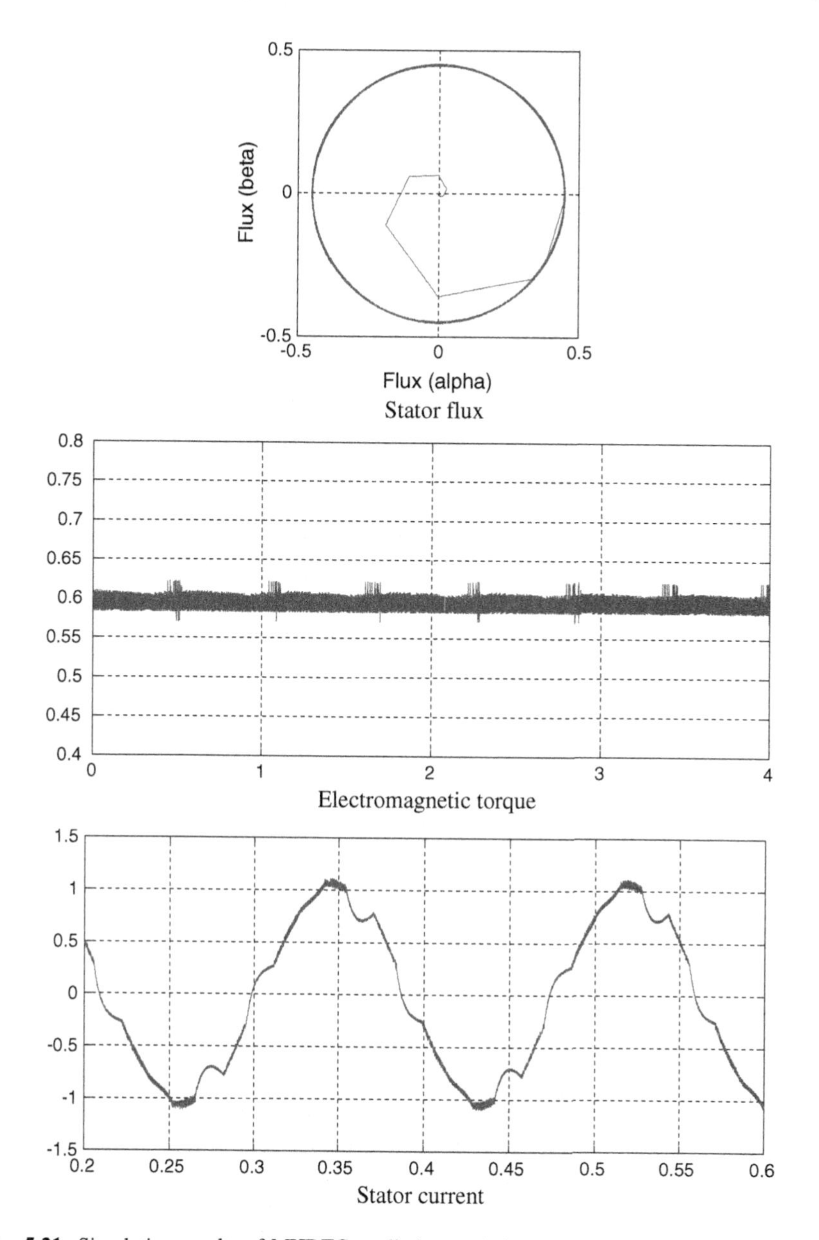

Fig. 5.21 Simulation results of MHDTC applied to an induction generator

where V_d and V_q are the Park transform of the three-phase stator voltages (Fig. 5.19).

We make application under MATLAB/SIMULINK. The model of power control is given as (Fig. 5.20):

Some simulation results can be represented (Fig. 5.21):

This method uses the advantages of the classical DTC and the FOC while correcting some of their difficulties. It leads to the following:

- a sinusoidal grid current with constant periodic frequency,
- a unit power factor, and
- a constant DC bus voltage.

5.3.5 Direct Power Control of Wind System

5.3.5.1 Direct Power Control Principals

In this control strategy, the error in the reference power and the actual power is utilized to generate the voltage control directly as in conventional DTC drives [12]. This method reduces the number of PI controllers used when compared to the vector control-based variable speed wind turbine generator systems. The direct power control (DPC) like DTC is a stator flux-based control technique having the advantages of robustness and fast controls.

5.3.5.2 DPC Structure

In the DPC structure, we use a PI controller in the DC link; this allows us to reduce the DC link capacitor fluctuation voltages.

5.3.5.3 Application: DPC of IM Fed by a Wind Turbine

Figure 5.22 shows the configuration of the proposed control system based on DPC method. The controller features relay control of the active and reactive power by using hysteresis comparators and a switching table. In this configuration, the DC bus voltage is regulated by adjusting the active power transmitted to the load. As shown in Fig. 5.22, the active power control P_{ref} is provided from the PI regulator of DC voltage controller block. The reactive power control Q_{ref} is directly given from the outside of the controller. Errors between the controlled and the estimated feedback power are input to the hysteresis comparators and digitized to the signals ΔP and ΔQ defined as [12–19]:

$$\begin{aligned} \Delta P = 1 \quad \text{if} \quad P_{\text{ref}} - P \geq \Delta h_p, \quad \Delta P = 0 \quad \text{if} \quad P_{\text{ref}} - P \leq -\Delta h_p \\ \Delta Q = 1 \quad \text{if} \quad Q_{\text{ref}} - Q \geq \Delta h_q, \quad \Delta Q = 0 \quad \text{if} \quad Q_{\text{ref}} - Q \leq -\Delta h_q \end{aligned} \tag{5.54}$$

where Δh_p and Δh_q are the hysteresis band (Fig. 5.23).

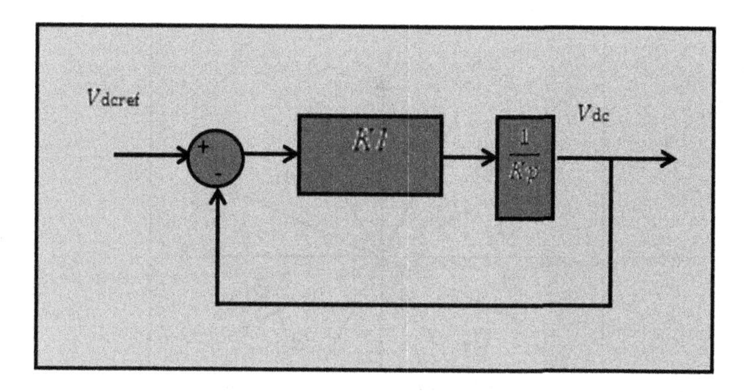

Fig. 5.22 PI controller

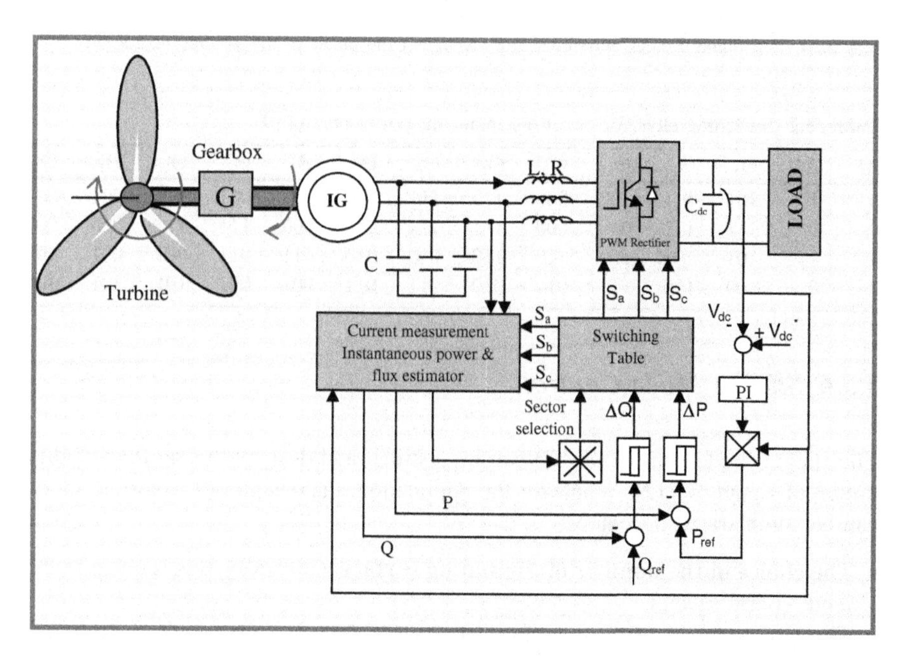

Fig. 5.23 Block scheme of DPC [9]

Also, the phase of the flux vector is converted to the digitized signals θ_n. For this purpose, the stationary coordinates are divided into twelve (12) sectors, as shown in Fig. 5.24, and the sectors can be numerically expressed as:

The digitized variables ΔP, ΔQ, and the flux vector position $\gamma_\psi = \text{arctg}(\psi_\beta/\psi_\alpha)$ form a digital word, which by accessing the address of the lookup table selects the appropriate voltage vector according to the switching Table 5.2.

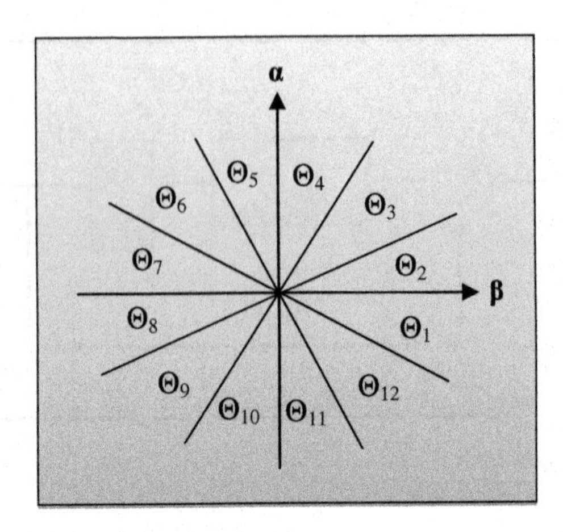

Fig. 5.24 α–β plane divided into 12 sectors to detect the phase of the voltage vector [9]

Table. 5.2 Switching table for direct instantaneous power control [9]

ΔP	ΔQ	Θ_1	Θ_2	Θ_3	Θ_4	Θ_5	Θ_6	Θ_7	Θ_8	Θ_9	Θ_{10}	Θ_{11}	Θ_{12}
0	1	101	111	100	000	110	111	010	000	011	111	001	000
0	0	111	111	000	000	111	111	000	000	111	111	000	000
1	1	101	100	100	110	110	010	010	011	011	001	001	101
1	0	100	110	110	010	010	011	011	001	001	101	101	100

5.3.6 Sliding Mode Control

The sliding mode control (SMC) can be justified and designed using the stability notion of Lypunov [13]. Whatever the application, the design of the sliding mode can be summarized in 3 steps:

- **The choice of the number of the sliding surface**. Generally, this is equal to the input control vector.
- **The choice of the sliding surface equation form**. It must satisfy the convergence of the control and the stability of the system. This goal can be reached if the control variable U_c permits to satisfy the Lyapunov function $S(x)\dot{S}(x) < 0$. Based on this condition, Slotine proposes a general form of sliding-mode surface [13]:

$$S(x) = \left(\frac{\delta}{\mathrm{d}t} + \lambda_x\right)^{r-1} (x^* - x) \tag{5.55}$$

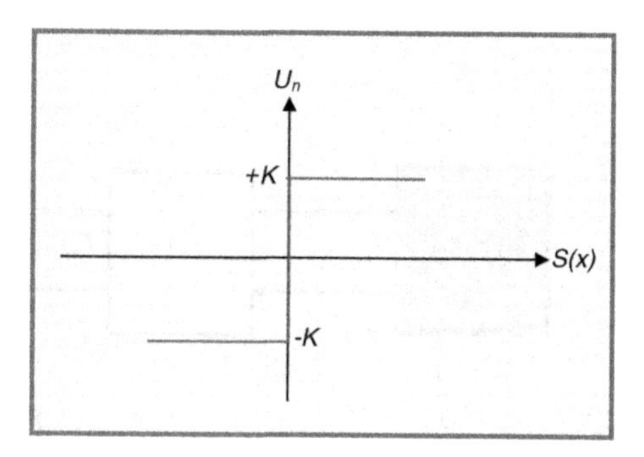

Fig. 5.25 Relay function

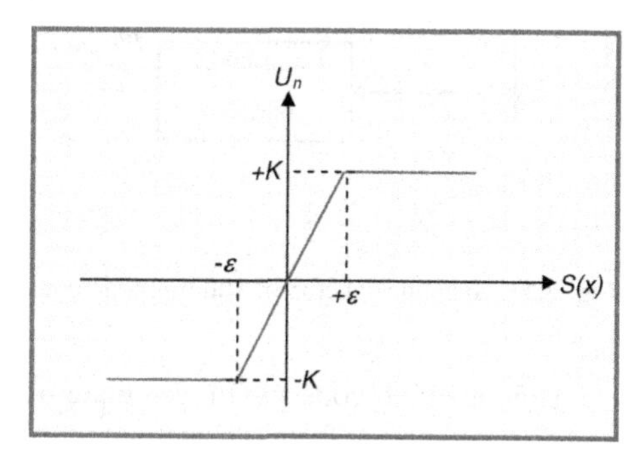

Fig. 5.26 Smoothed sign function

- **The control law design**. The control variable is decomposed into two parts: U_{eq} and U_n:

$$U_c = U_{eq} + U_n \tag{5.56}$$

The dynamic while in sliding mode can be written as: $\dot{S} = 0$. By solving this equation, the equivalent control U_{eq} can be obtained [13].

The nonlinear component U_n satisfies $S(x)\dot{S}(x) < 0$ and is given by:

$$U_n = K \cdot \text{sign}(S(x)) \tag{5.57}$$

The common used form of U_n is a constant relay control (Fig. 5.25) [13].

However, this latter produces a drawback in the performances of a control system, which is known as chattering phenomenon. In order to reduce the chattering phenomenon due to the discontinuous nature of the controller, a smooth function is defined in some neighborhood of the sliding surface with a threshold (Fig. 5.26) [13].

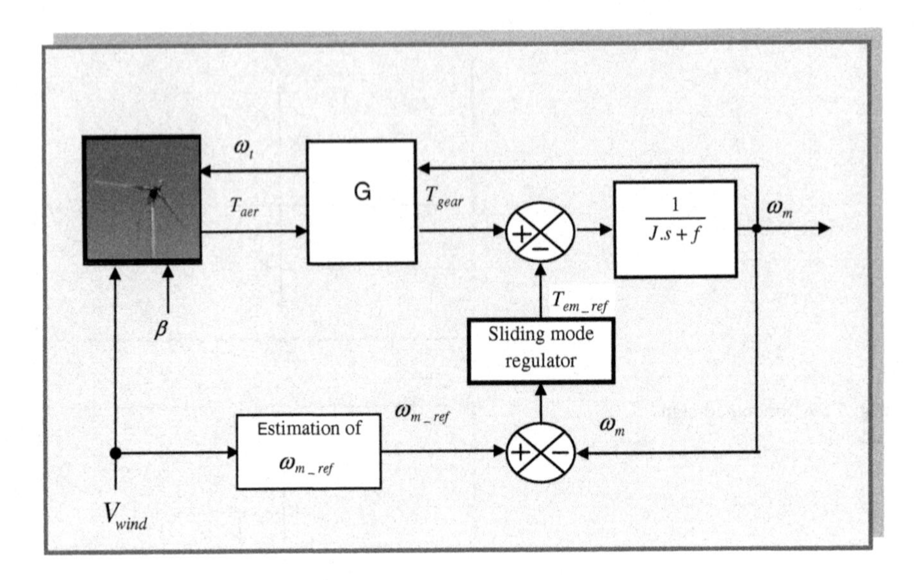

Fig. 5.27 Synoptic scheme of sliding-mode control strategy

This method consists to calculate the two control components (equivalent and discontinuous) from an adequate surface [13]. In this case, we chose the error between the variables of control like a simple surface of sliding mode. For the mechanical speed, the surface is given by

$$S(\omega_m) = \omega_m - \omega_{m\_\mathrm{ref}} \tag{5.58}$$

The first derivate of Eq. 5.58, gives:

$$\dot{S}(\omega) = \dot{\omega} - \dot{\omega}_{\mathrm{ref}} \tag{5.59}$$

In the sliding-mode regime arise, the dynamic of the system in sliding mode is subjected to the following equation $S(\psi) = 0$; thus, for the ideal sliding mode, we have also $\dot{S}(\psi) = 0$ [13].

We replace Eq. 5.59 and with the conditions of the sliding mode; we obtain:

$$T_{\mathrm{em\_eq}} = T_{\mathrm{gear}} - J \cdot \dot{\omega}_{\mathrm{ref}} \tag{5.60}$$

where $T_{\mathrm{em\_eq}}$ is the equivalent component of the reference electromagnetic torque.

The reference electromagnetic torque is defined by:

$$T_{\mathrm{em\_ref}} = T_{\mathrm{em\_eq}} + T_{\mathrm{em}\_n} \tag{5.61}$$

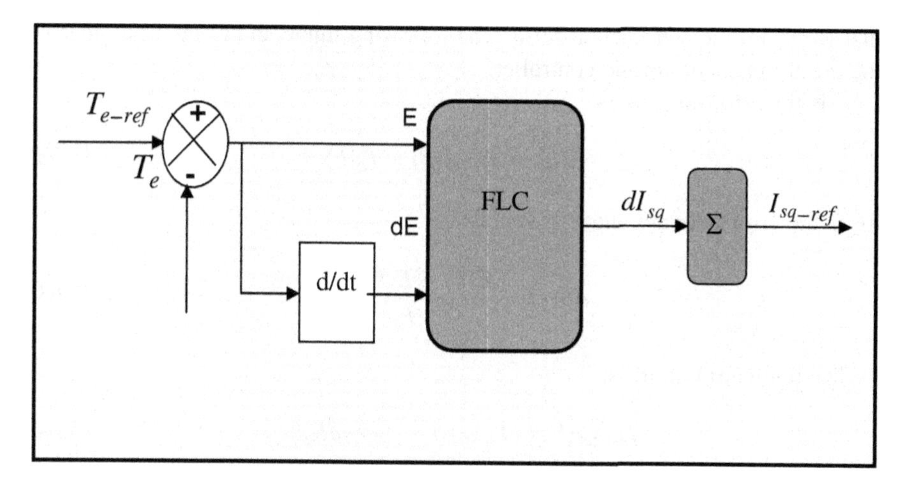

Fig. 5.28 Basic scheme of electromagnetic torque fuzzy controller

Table 5.3 Fuzzy rules table

dI_{sq}		dE				
		GN	PN	Z	PP	GP
E	GN	GN	GN	PN	PN	Z
	PN	GN	PN	PN	Z	PP
	Z	GN	PN	Z	PP	GP
	PP	PN	Z	PP	PP	GP
	GP	Z	PP	PP	GP	GP

The nonlinear component of electromagnetic torque T_{em_n} satisfies the Lypunov condition $S(x)\dot{S}(x) < 0$ and is given by a relay or smoothed function (Figs. 5.25, 5.26 and 5.27).

The synoptic scheme of the SMC strategy is given by Fig. 5.28.

5.3.7 *Fuzzy Logic Controller*

Fuzzy logic controller (FLC) is explained in Sect. 5.3.5. It was applied to search maximum power point.

5.3.7.1 Fuzzy Logic Controller Applied to a WECS

Classic PI controllers are replaced by fuzzy controllers for controlling the stator flow and the electromagnetic torque of the electrical generator. These controllers are based on the same structure of a conventional controller with the difference that we retain the incremental form. The following figure shows the basic idea of the electromagnetic torque controller.

E is the error defined as:

$$E(k) = T_{\text{e-ref}}(k) - T_e(k) \quad (5.62)$$

dE is the error derivate and is given as:

$$dE(k) = \frac{E(k) - E(k-1)}{T_e} \quad (5.63)$$

The regulator output is:

$$I_{\text{sq-ref}}(k) = I_{\text{sq-ref}}(k-1) + dI_{\text{sq}}(k) \quad (5.64)$$

The fuzzy rules used are listed in Table 5.3.

5.4 Conclusion

Several nonlinear controls of wind turbine systems have been presented in this chapter. The mechanical and the electrical levels have been studied separately. We have presented three applications under MATLAB/SIMULINK of the NLCSSF, the NLDCSF, and the indirect speed control. From the studied example, the most effective strategy is the NLDCSF. It rejects the disturbance on the control and achieves better tracking of the optimal speed.

References

1. Noguchi T, Tomiki H, Kondo S, Takahashi I (1998) Direct power control of PWM converter without power–source voltage sensors. IEEE Trans Ind Appl 34:473–479
2. Ohnishi T (1991) Three phase PWM converter/inverter by means of instantaneous active and reactive power control. In: Proceedings of IEEE IECON, pp 819–824
3. Bouafia A, Krim F, Gaubert JP (2009) Design and implementation of high performance direct power control of three-phase PWM rectifier, via fuzzy and PI controller for output voltage regulation. Energy Convers Manage 50(1):6–13
4. Hansen S, Malinowski M, Blaabjerg F, Kazmierkowski MP (2000) Control strategies for PWM rectifiers without line voltage sensors. In: Proceedings of IEEE APEC 2:832–839
5. Abdelli R, Rekioua D, Rekioua T, Tounzi A (2013) Improved direct torque control of an induction generator used in a wind conversion system connected to the grid. ISA Trans 52(4):525–538
6. Malinowski M, Kazmierkowski MP, Hansen S, Blaabjerg F, Marques GD (2001) Virtual flux based direct power control of three-phase PWM rectifiers. IEEE Trans Ind Appl 37(04):1019–1027

7. Rekioua D, Matagne E (2012) Optimization of photovoltaic power systems: modelization, simulation and control. Green Energy and Technol 102
8. Malinowski M, Kazmierkowski MP (2002) Direct power control of three-phase PWM rectifier using space vector modulation-simulation study. In: Proceedings of IEEE-ISIE 4: 1114–1118
9. Boudries Z, Rekioua D (2013) Study on decoupling direct power control of PWM rectifier using space vector modulation (4):875–882
10. Beltran B, Ahmed-Ali T, Benbouzid MEH (2007) Sliding mode power control of variable speed wind energy conversion systems. In: IEEE international electric machines and drives conference IEMDC (vol 2)
11. Rekioua D, Rekioua T, Idjdarene K, Tounzi A (2005) An approach for the modeling of an autonomous induction generator taking into account the saturation effect. Int J Emerg Electr Power Syst 4(1):1–25
12. Mcgowan JG, Manwell JF (1999) Hybrid/pv/diesel system experiences. Rev Renew Energy 16:928–933
13. Belhamel M, Moussa S, Kaabeche A (2002) Production of electricity by means of a Hybrid System (Wind-Photovoltaic-Diesel). Rev Renew Energy, pp 49–54
14. Rekioua D, Bensmail S, Bettar N (2014) Development of hybrid photovoltaic-fuel cell system for stand-alone application. Int J Hydrogen Energy 39(3):1604–1611
15. Kaldellisa JK, Kavadiasa KA, Koronakis PS (2007) Comparing wind and photovoltaic stand-alone power systems used for the electrification of remote consumers. Renew Sustain Energy Rev 11:57–77
16. El Khadimi A, Bachir L, Zeroual A (2004) Sizing optimization and techno-economic energy system hybrid photovoltaic—wind with storage system. Renew Energy J 7:73–83
17. Koussa D, Alem M, Belhamel M (2002) Hybrid system (wind, solar) for the power supply for a load for household. J Renew Energy p 1–8
18. Kaldellisa JK, Kavadiasa KA, Koronakis PS (2007) Comparing Wind And Photovoltaic Stand-Alone Power Systems Used For The Electrification of Remote Consumers. Renew Sustain Energy Rev 11:57–77
19. Vechiu I (2005) Modelling and analysis of integration of renewable energy in an autonomous network. PhD Thesis, University of Havre, France

Chapter 6
Hybrid Wind Systems

6.1 Advantages and Disadvantages of a Hybrid System

6.1.1 *Advantages of Hybrid System*

- Not dependent on one source of energy.
- Simple to use.
- Efficiency, low cycle cost of living component of the hybrid system.

6.1.2 *Disadvantages of a Hybrid System*

- More complex than single-source systems and the need for storage,
- high capital cost compared to diesel generators.

6.2 Configuration of Hybrid System

Photovoltaic and wind generators in a hybrid system can be connected in three configurations, DC bus architecture, AC bus architecture, and DC–AC bus architecture [1, 2].

6.2.1 *Architecture of DC Bus*

In the hybrid system presented in Fig. 6.1, the power supplied by each source is centralized on a DC bus. Thus, the energy conversion system to provide AC power at their first rectifier has to be converted then continuously. The generators are connected in series with the inverter to power the load alternatives. The inverter

D. Rekioua, *Wind Power Electric Systems*, Green Energy and Technology,
DOI: 10.1007/978-1-4471-6425-8_6, © Springer-Verlag London 2014

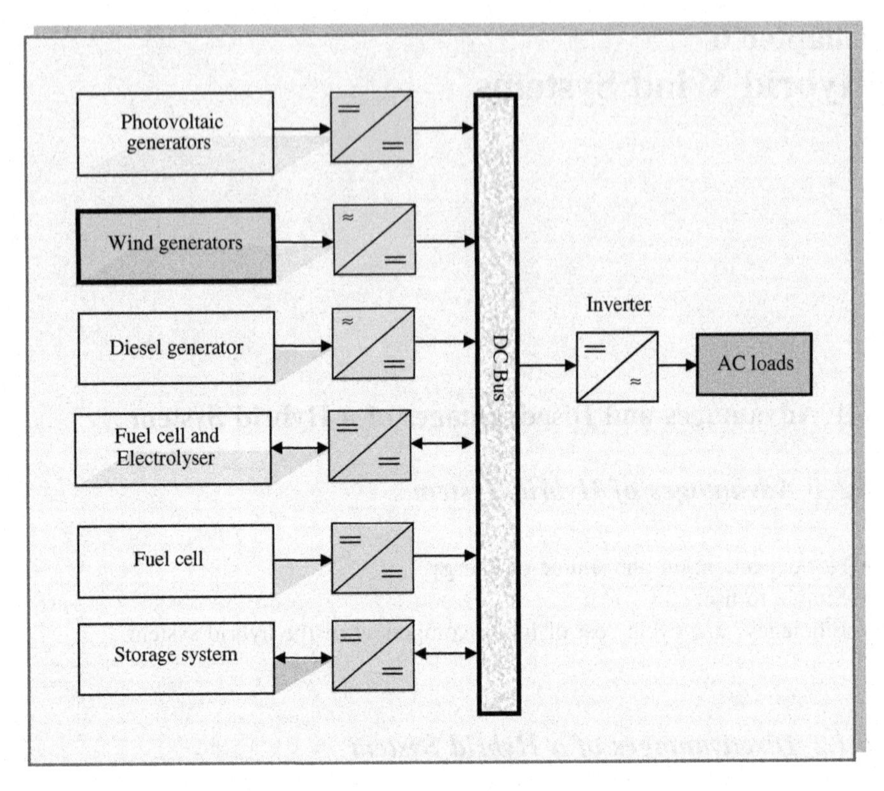

Fig. 6.1 Configuration of the hybrid system with DC bus

should supply the alternating loads from the DC bus and must follow the set point for the amplitude and frequency [3]. The batteries are sized to supply peak loads. The advantage of this topology is the simplicity of operation, and the load demand is satisfied without interruption even when the generators charge the short-term storage units.

6.2.2 Architecture of AC Bus

In this topology, all components of the HPS are related to alternating loads, as shown in Fig. 6.2. This configuration provides superior performance compared to the previous configuration, since each converter can be synchronized with the generator so that it can supply the load independently and simultaneously with other converters [1]. This provides flexibility for the energy sources which fed the load demand. In the case of low load demand, all generators and storage systems

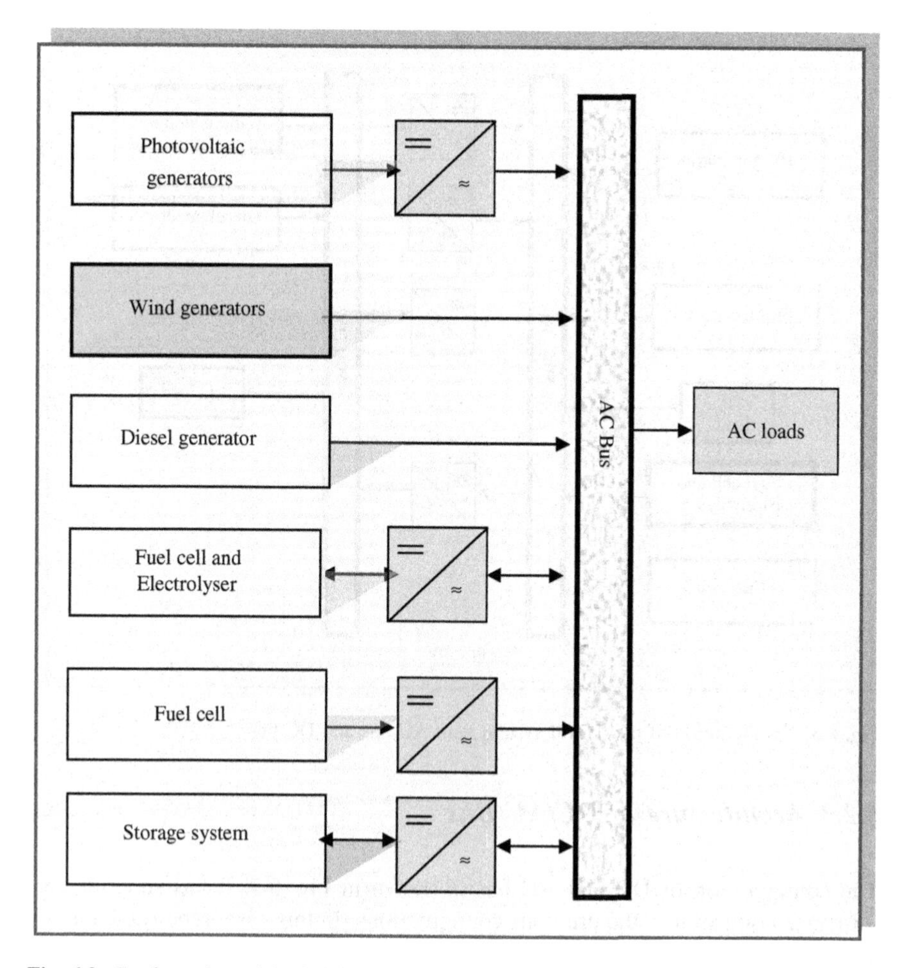

Fig. 6.2 Configuration of the hybrid system with AC bus

are stationary except, for example, the photovoltaic generator to cover the load demand. However, during heavy load demands or during peak hours, generators and storage units operate in parallel to cover the load demand. The realization of this system is relatively complicated because of parallel operation, by synchronizing the output voltages with the charge voltages [2]. This topology has several advantages compared to the DC coupled topology such as higher overall efficiency, smaller sizes of the power conditioning unit while keeping a high level of energy availability, and optimal operation of the diesel generator due to reducing its operating time and consequently its maintenance cost [4].

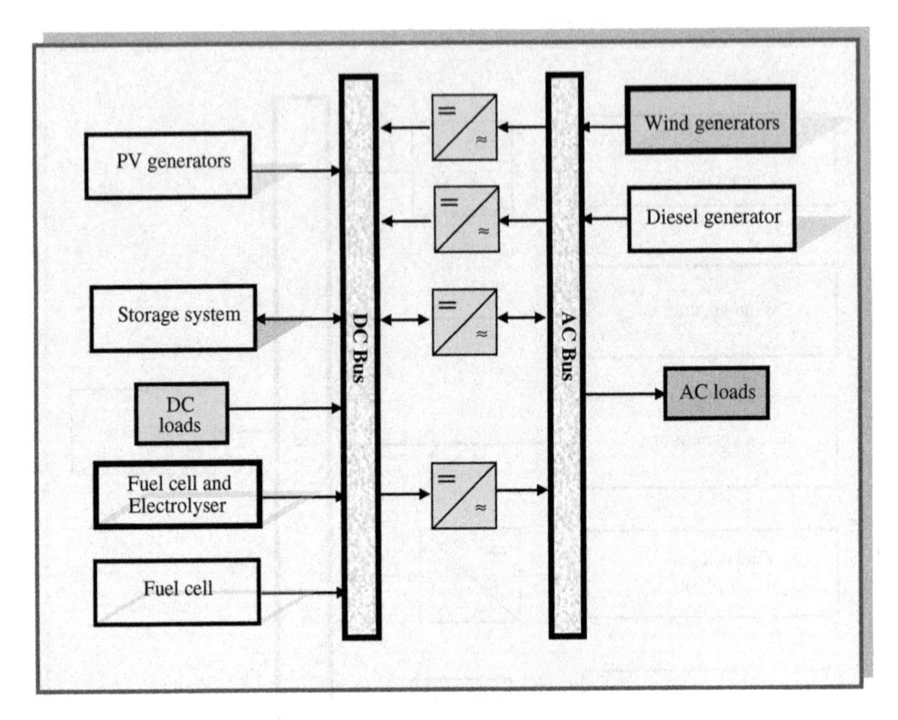

Fig. 6.3 Configuration of the hybrid system with AC bus and DC bus

6.2.3 Architecture of DC/AC Bus

The configuration of DC and AC bus is shown in Fig. 6.3. It has superior performance compared to the previous configurations. In this case, renewable energy and diesel generators can power a portion of the load directly to AC, which can increase system performance and reduce power rating of the diesel generator and the inverter. The diesel generator and the inverter can operate independently or in parallel by synchronizing their output voltages. Converters located between two buses (the rectifier and inverter) can be replaced by a bidirectional converter, which in normal operation, performs the conversion DC/AC (inverter operation). When there is a surplus of energy from the diesel generator, it can also charge batteries (operating as a rectifier). The bidirectional inverter can supply the peak load when the diesel generator is overloaded [5].

The advantages of this configuration are:

- The diesel generator and the inverter can operate independently or in parallel. When the load level is low, one or the other can generate the necessary energy. However, both sources can operate in parallel during peak load.
- The possibility of reducing the nominal power of the diesel generator and the inverter without affecting the system's ability to supply peak loads.

Table 6.1 Classification of hybrid systems by power range

Power of hybrid system [kW]	Applications
Low power: <5	Autonomous system: telecommunication stations, pumping water, other isolated applications
Average power: 10–250	Microisolated systems: feeding a remote village, rural
Great power: >500	Large isolated systems (Islands)

The disadvantages of this configuration are:

- The implementation of this system is relatively complicated because of the parallel operation (the inverter should be able to operate autonomously and operate with synchronization of the output voltages with output voltages of diesel generator).

6.2.4 Classifications of Hybrid Energy Systems

The power delivered by the hybrid system can vary from a few watts for domestic applications up to a few megawatts for systems used in the electrification of small islands [6]. Thus, for hybrid systems with a power below 100 kW, the configuration with AC and DC bus, with battery storage, is the most used. The storage system uses a high number of batteries to be able to cover the average load for several days. This type of hybrid system uses small renewable energy sources connected to the DC bus. Another possibility is to convert the continuous power to an alternative one by using inverters. Hybrid systems used for applications with very low power (below 5 kW) supply generally DC loads (Table 6.1).

6.3 Different Combinations of Hybrid Systems

6.3.1 Hybrid Wind/Photovoltaic System

The optimization of wind energy, photovoltaic with electrochemical storage (batteries), depends on many economic models of each system separately (wind and photovoltaic). The advantage of a hybrid system depends on many important factors: the shape and type of load, the wind, solar radiation, cost and availability of energy, the relative cost of the machine wind, solar array, electrochemical storage system, and other efficiency factors [7]. Photovoltaic systems are currently economical for low power installations. For autonomous systems, the cost of energy storage is the biggest constraint the overall system cost for large power

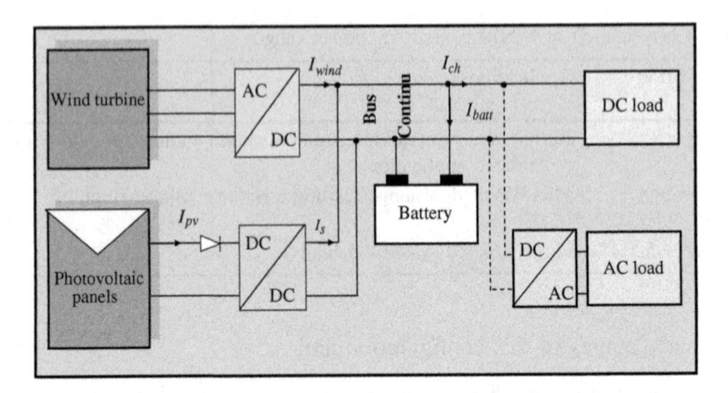

Fig. 6.4 Hybrid wind/photovoltaic system

installations. Minimizing the cost of storage and reducing its capacity are the main reasons for the combination of wind and photovoltaic systems [8]. This type of hybrid system includes a photovoltaic subsystem. A DC/DC parallel type can catch up whenever the maximum power point, for example, a wind turbine that converts wind energy into electricity. Both energy sources are connected to a DC bus. Battery and inverter are included as part of backup and storage system (Fig. 6.4).

6.3.2 *Sizing of Hybrid Wind/Photovoltaic System*

The effectiveness of any electric system depends on its sizing and its use. The sizing should be based on meteorological data, solar radiation, and wind speed; and the exact load profile of consumers over long periods.

6.3.2.1 Determination of the Load Profile of Consumers

The exact knowledge of the customers load profile determines the size of generator [9].

6.3.2.2 Analysis of Solar and Wind Energy Potential

We make application in Bejaia (Algeria) which is a coastal region with two complementary sources (wind speed and radiation), so the coupling of a photovoltaic system and wind is very interesting for the production electricity throughout the year.

6.3.2.3 Photovoltaic Energy Calculation

The energy produced by a photovoltaic generator is estimated using data from the global irradiation on an inclined plane, ambient temperature, and the data sheet used for the photovoltaic panel.

The electrical energy produced by a photovoltaic generator is given by:

$$E_{\mathrm{pv}} = \eta_{\mathrm{pv}} \cdot A_{\mathrm{pv}} E_{\mathrm{s}} \tag{6.1}$$

with A_{pv}: the total area of the photovoltaic generator; η_{gen}: the efficiency of the photovoltaic generator.

$$\eta_{\mathrm{pv}} = \eta_{\mathrm{r}} \cdot \eta_{\mathrm{pc}} \left[1 - \alpha_{\mathrm{sc}} \left(T_j - T_{\mathrm{jref}}\right)\right] \tag{6.2}$$

where E_{s} is a solar radiation on tilted plane module, η_{r} is the reference efficiency of the photovoltaic generator, η_{pc} is the power conditioning efficiency which is equal to 1 if a perfect maximum power tracker (MPPT) is used, α_{sc} is the temperature coefficient of short current (A/°K) and found on the data sheet, T_{j} cell temperature, T_{jref} is the reference cell.

6.3.2.4 Wind Energy Calculation

The power contained in the form of kinetic energy, in the wind, is expressed by:

$$P_{\mathrm{wind}} = \frac{1}{2} \cdot \rho \cdot S_{\mathrm{wind}} \cdot v_{\mathrm{wind}}^{3}. \tag{6.3}$$

The energy produced by wind generator is expressed by:

$$E_{\mathrm{wind}} = P_{\mathrm{wind}} \cdot \Delta t. \tag{6.4}$$

6.3.2.5 Sizing of Photovoltaic and Wind Systems

The monthly energy produced by the system per unit of area is denoted $E_{\mathrm{pv,m}}$ (kWh/m^2) for photovoltaic energy and $E_{\mathrm{wind,m}}$ (kWh/m^2)^2 for wind energy and $E_{\mathrm{L,m}}$ represents the energy required by load every month (where m = 1, 2, …, 12 represents the month of the year).

The total energy produced by both photovoltaic and wind a generator supplying the load is expressed by:

$$E_{\mathrm{L}} = E_{\mathrm{pv}} \cdot A_{\mathrm{pv}} + E_{\mathrm{wind}} \cdot S_{\mathrm{wind}} \tag{6.5}$$

where

- For photovoltaic generator

$$A_{pv} = \frac{E_{L,m}}{E_{pv,m}}. \tag{6.5}$$

- For wind generator

$$S_{wind} = \frac{E_{L,m}}{E_{wind,m}} \tag{6.6}$$

with

$$\begin{aligned} E_{pv} \cdot A_{pv} &= f_{pv} \cdot E_L \\ E_{wind} \cdot S_{wind} &= \left(1 - f_{pv}\right) \cdot E_L = f_{wind} \cdot E_L \end{aligned} \tag{6.7}$$

where f_{pv} is the fraction of load supplied by the photovoltaic energy.

- $f_{wind} = (1 - f_{pv})$ is the fraction of load supplied by the wind energy.
- $f_{pv} = 1$: indicates that the entire load is supplied by the photovoltaic source.
- $f_{pv} = 0$: indicates that the entire load is powered by the wind source.

The sizing is based on monthly annual average [10, 11]. The calculation of the size of wind generator and photovoltaic (A_{pv} and S_{wind}) is established from the annual average values of each monthly contribution $\left(\overline{E_{pv}} \text{ and } \overline{E_{wind}}\right)$. The load is represented by the annual average energy $\overline{E_L}$.

$$\begin{aligned} A_{pv} &= f \cdot \frac{\overline{E_L}}{\overline{E_{pv}}} \\ S_{wind} &= (1 - f) \cdot \frac{\overline{E_L}}{\overline{E_{wind}}} \end{aligned} \tag{6.8}$$

The number of photovoltaic and wind generator to consider is calculated according to the area of the system unit taking the full value of the report by excess.

$$\begin{aligned} N_{pv} &= \text{ENT}\left[\frac{A_{pv}}{A_{pv,u}}\right] \\ N_{wind} &= \text{ENT}\left[\frac{S_{wind}}{S_{wind,u}}\right] \end{aligned} \tag{6.9}$$

6.3.2.6 Sizing Batteries

(see Chap. 1), Sect. 1.3.5.

6.3.2.7 Control of Hybrid Photovoltaic/Wind System

Managing energy sources (photovoltaic and wind) is provided by a supervisor. For the design of the supervisor, it was decided that the subphotovoltaic system would be the main generator, while the subsystem wind generator would be complementary. This choice is motivated by the design already made based on monthly averages annual rating site. However, the supervisor applications extend to considering the subwind system as the main generator, and the photovoltaic subsystem would be complementary.

Three operating modes are possible to determine the ability of the hybrid system to supply the total power required (the power load and the power required to charge the batteries) and those based on atmospheric conditions (insolation, temperature, and wind speed). This supervisor is essential to effectively control energy subsystems (photovoltaic, wind). We can have three cases [9, 12].

- **Case 1**

This mode corresponds to the periods or photovoltaic power is sufficient to supply the load demand. However, the PV generator must provide the total power while the wind subsystem is supposed stop and the batteries are charging. This situation is maintained until the power required by the load does not exceed the maximum PV power. Beyond this limit, the supervisor switches in the case 2 and activates the wind generator. In this case, the objective of the photovoltaic system is under power control according to this reference:

$$P_{\text{ref1\_PV}} = P_{\text{required}} = V_{\text{batt}} \cdot (I_{\text{load}} + I_{\text{batt}}). \tag{6.10}$$

With I_{load}: the load's current, I_{batt}: the battery's current, P_{required}: the total required power.

- **Case 2**

In this case, the photovoltaic system generates the maximum power (operating at maximum power point ($\text{MPPT}_{\text{w}} = 1$), and the wind system is controlled to produce a reference power. This one is the power required to complete the power produced by the photovoltaic generator at the same time supplying the total power load. It should be noted that in cases 1 and 2, batteries are not used to produce load power, instead they become a part of the power required. Once the maximum production limit of the hybrid system is reached or exceeded by any power demand, the system switches in the case 3.

In the cases 2 and 3, the PV system produces maximum power at MPPT operation. Different algorithms can be used to extract the maximum power (see Chap. 4). The reference power is given by:

$$P_{\text{ref2\_PV}} = P_{\text{pv}}^{\text{opt}} = P_{\text{s}}^{\text{opt}} = V_{\text{pv}}^{\text{opt}} \cdot I_{\text{pv}}^{\text{opt}} \tag{6.11}$$

The wind system starts its operation when the PV power is insufficient to supply the total power required. The supervisor controls the wind system by power control or by maximum power operation. The objective in case 2 is to produce the additional power to supply the total power applied. The wind power reference is given by:

$$P_{\text{ref1\_w}} = P_{\text{required}} - P_{\text{s}}^{\text{opt}} = V_{\text{batt}} \cdot (I_{\text{load}} + I_{\text{batt}} - I_{\text{s}}) \tag{6.12}$$

When the contribution of wind power subsystem is no longer sufficient to supply the total power required, the supervisor switches in the case 3. The objective of this subsystem is the generation of maximum power extraction.

- **Case 3**

In this case, the two photovoltaic and wind generators provide maximum power (operating at MPPT). In addition, to supply the load demand, the batteries are charged or discharged. At discharge, the case 3 is maintained as long as the available energy levels of the batteries is sufficient to complete the load demand, after that, the load must be disconnected to charge the batteries. The wind system produces maximum power MPPT, the reference power is given by:

$$P_{\text{ref2\_w}} = P_{\text{w}}^{\text{opt}} = K_{\text{opt}} \cdot \omega_{\text{opt}}^{3} \tag{6.13}$$

with K_{opt}: a coefficient which depends on the ratio of tip speed to optimal power coefficient maximum. The reference angular velocity which corresponds to the operating MPPT is given by:

$$\omega_{\text{ref}} = \omega_{\text{opt}} = \sqrt{[3]}\frac{P_{\text{ref2-w}}}{K_{\text{opt}}} \tag{6.14}$$

Then, the supervisor decides the case (1 or 2/3) by comparing the measured mechanical speed with the reference speed.

$$\begin{cases} \text{If} \quad \omega < \omega_{\text{opt}}, & \text{case 1}, \quad P_{\text{w}} = P_{\text{ref1\_w}} \\ \text{If} \quad \omega = \omega_{\text{opt}}, & \text{case 2/3}, \quad P_{\text{w}} = P_{\text{ref2\_w}} = P_{\text{w}}^{\text{opt}} \end{cases} \tag{6.15}$$

A description of operating cases is shown in Fig. 6.5.

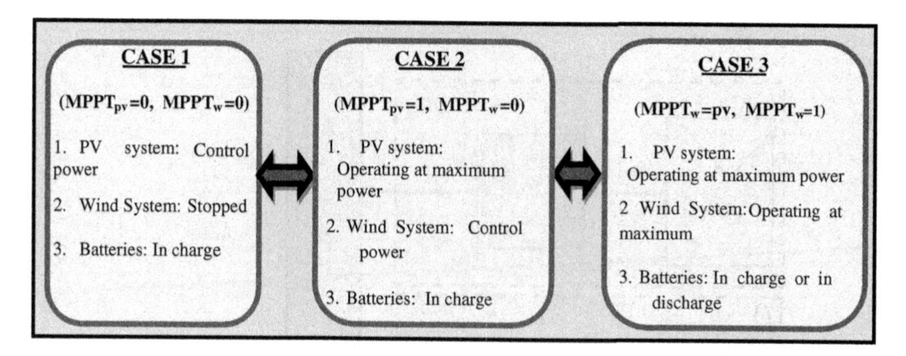

Fig. 6.5 Description of operating cases

6.4 Hybrid Wind/Photovoltaic/Diesel Generator System

This type of hybrid system is well suited for decentralized production of electricity and can contribute to solving the problem of connecting to the electricity networks (cases of isolated sites) [6, 13]. The initial data in the implementation of such a system of production to renewable sources of energy like any other energy system are the demand, which will be determined by comparing the load to be supplied. This request must be estimated as accurately as possible both from a standpoint of powers called as its temporal distribution, even if its random nature makes this often difficult task. Adding a generator to a system of renewable energy production may on the one hand increase the reliability of power system loads and to reduce significantly the cost of electricity produced by a significant decrease of the size of the storage system [5, 7, 14].

Ref [11] proposed that there are multiple types of electrical circuit architectures which could be used depending on people needs and site capabilities. In the first architecture (Fig. 6.6), the generators and the battery are all installed in one place and are connected to a main AC bus bar before being connected to the grid. The power delivered by all the energy conversion systems and the battery is fed to the grid through a single point. In this case, the power produced by the PV system and the battery is inverted into AC before being connected to the main AC bus. This system is called centralized AC bus architecture.

The energy conversion systems can also be connected to the grid in another manner (Fig. 6.7). This system is called decentralized AC bus architecture. The power sources in this case do not need to be connected to one main bus bar. The power generated by each source is conditioned separately to be identical to that required by the grid.

The third architecture uses a main centralized DC bus bar (Fig. 6.8). The energy conversion systems produced by AC power (wind energy converter and the diesel generator) deliver their power to rectifiers to be converted into DC, and then, it is delivered to the main DC bus bar. A main inverter feeds the AC grid from this main DC bus.

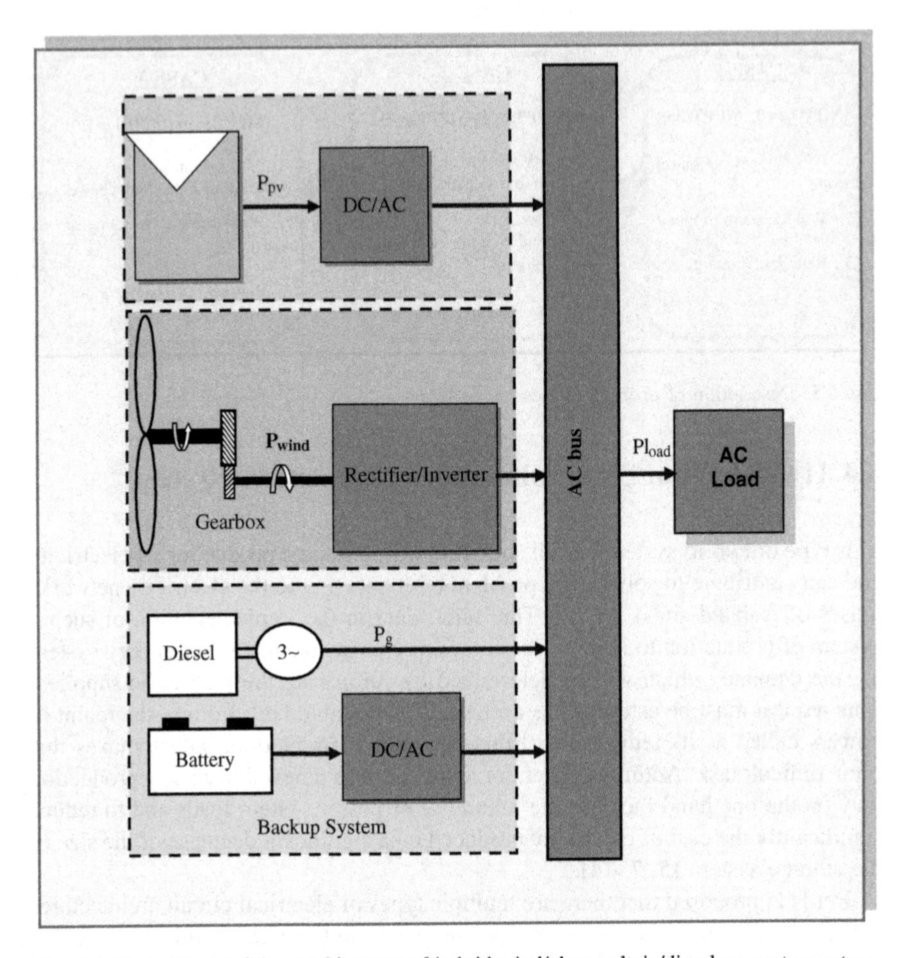

Fig. 6.6 Centralized AC bus architecture of hybrid wind/photovoltaic/diesel generator system

The monitoring equipment includes data loggers, wind speed and direction sensors, ambient and battery temperature sensors, and various AC and DC current/voltage/power sensors. The purposes for using monitoring systems are [15]:

- Determine components and system efficiencies.
- Verify proper system functioning.
- Provide system trouble shooting.
- Detect and analyze significant load changes.
- Calculate actual cost of utilized energy.

We propose a system control of the hybrid wind/PV//diesel system [16]. It is based on the overall energy balance equation.

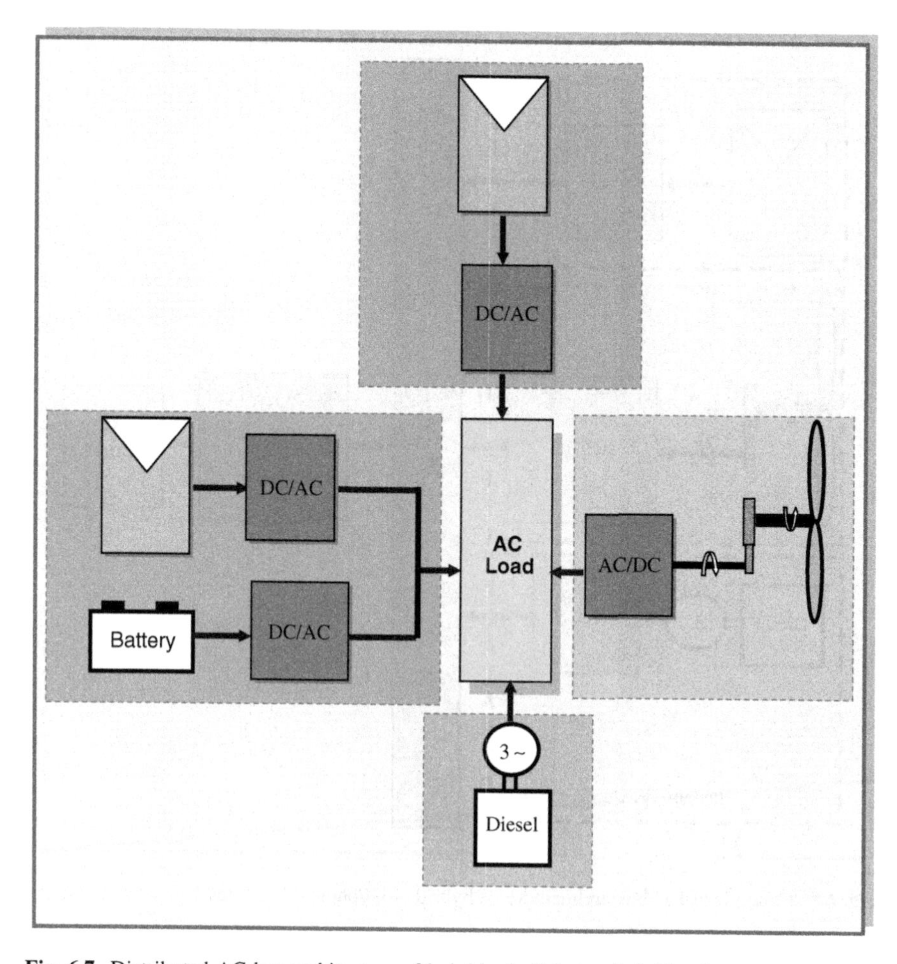

Fig. 6.7 Distributed AC bus architecture of hybrid wind/photovoltaic/diesel generator system

$$P_{\text{diesel}} = P_{\text{load}} - P_{\text{wind}} - P_{\text{pv}} + \Delta P_{\text{diss}} - P_{\text{unm}} \tag{6.16}$$

where P_{diesel} is the power delivered from the diesel generator(s), P_{load} is the power required by the load, P_{wind} is the power delivered from the wind turbine, ΔP_{diss} is the power dissipated in the dump load, P_{unm} is the unmet load, and P_{pv} is the power delivered from the PV.

The power control unit (PCU) is a central location for making the various connections of subsystems (wind, photovoltaic, diesel generator) (Fig. 6.9).

The monitoring system's role is to manage and control the operation of hybrid power system, depending on weather (irradiance, wind speed) and the power required. The manager controls the opening and closing of three relays under the following conditions:

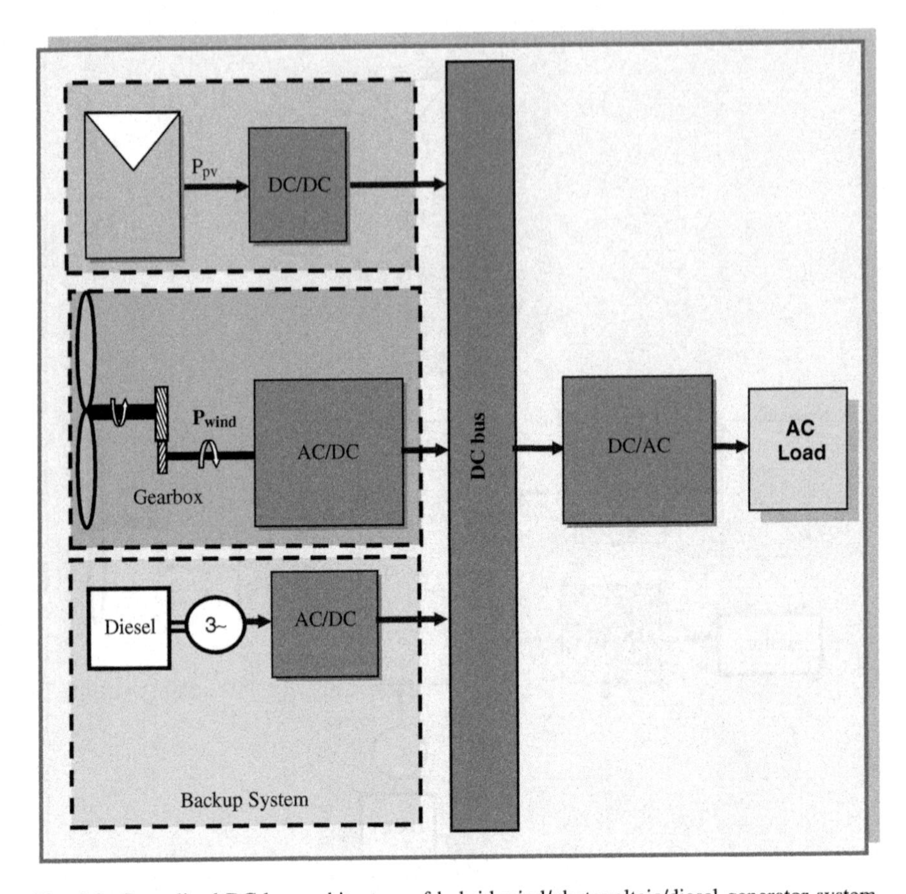

Fig. 6.8 Centralized DC bus architecture of hybrid wind/photovoltaic/diesel generator system

- **The relay of the PV generator is open if**:
  - batteries are charged
  - current output by the PV generator is zero
  - load power is zero.
- **The relay wind generator is open if**:
  - batteries are charged
  - wind speed is less than the initial speed of the turbine
  - wind speed is greater than the stall speed of the wind
  - load power is zero.
- **The relay of the diesel generator is open if**:
  - batteries are charged
  - generators (wind and PV) give a power greater than the load power
  - load power is zero, and the batteries are charged.

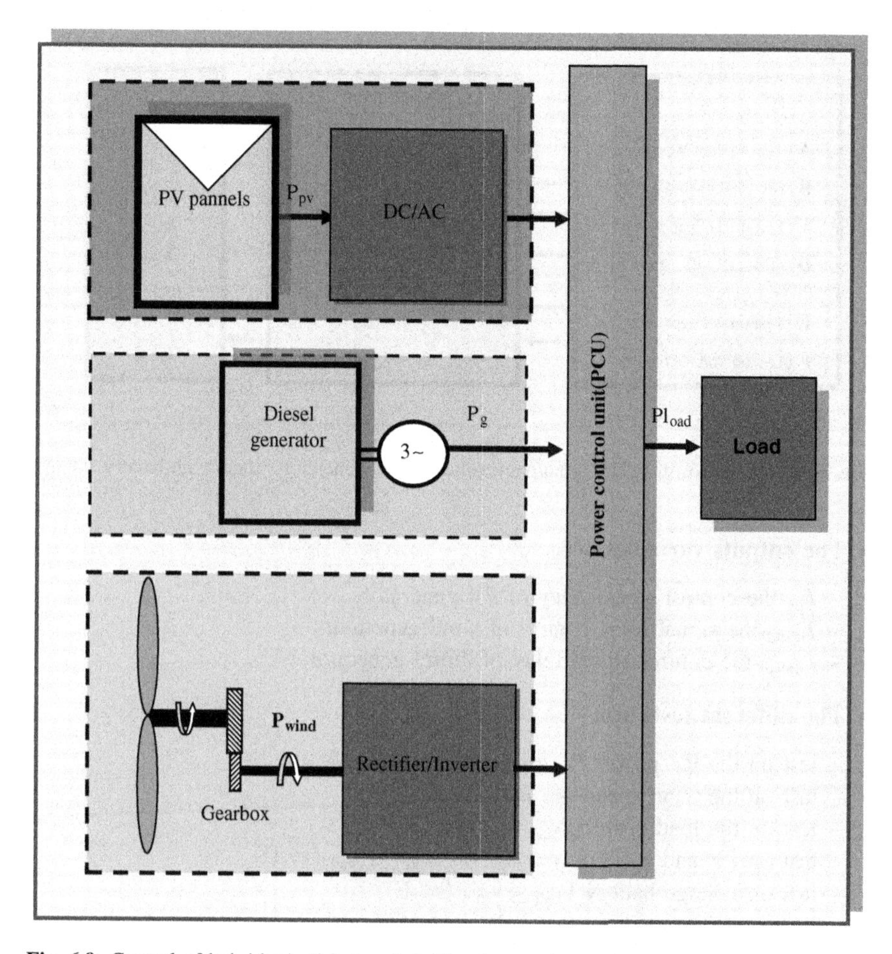

Fig. 6.9 Control of hybrid wind/photovoltaic/diesel generator system

And the closure of this relay is when the state of battery charge reaches the minimum level.

From these conditions, we find that the monitoring system includes 06 inputs, 03 outputs, and 06 tests.

- **The inputs variables are:**
  - insolation (G),
  - wind speed (V_{wind}),
  - power of PV generator (P_{pv}),
  - wind power generator (P_{wind}),
  - power load (P_{load}),
  - battery voltage (V_{batt}).

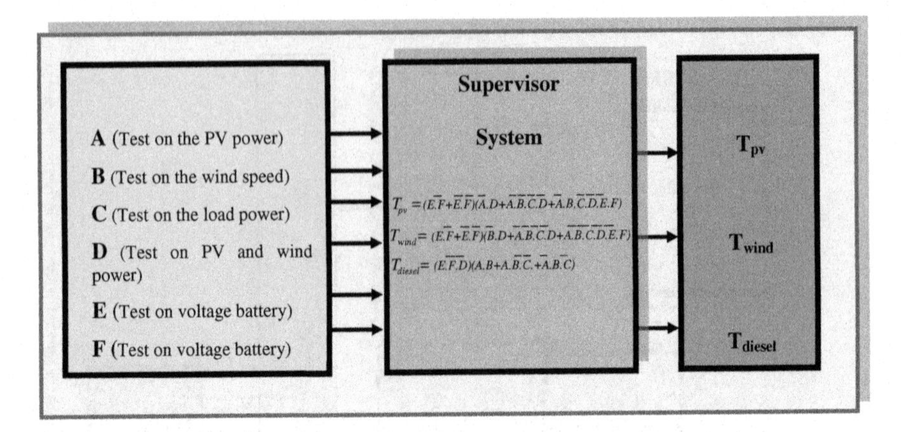

Fig. 6.10 Supervisor of hybrid wind/photovoltaic/diesel generator system with battery storage

- **The outputs variables are:**
  - T_{pv} the control signal relay of PV generator
  - T_{wind} the signal relay control of wind generator
  - T_{diesel} the control signal relay of diesel generator.
- **The different tests are:**
  - test on the PV power $P_{pv} = 0$ or $G = 0$ ($\Leftrightarrow$ A)
  - test on the wind speed ($\Leftrightarrow$ B)
  - test on the load power $P_{load} = 0$ ($\Leftrightarrow$ C)
  - test on PV and wind power $P_{pv} + P_{wind} \geq P_{load}$ ($\Leftrightarrow$ D)
  - test on voltage battery $V_{batt} \leq V_{min}$ ($\Leftrightarrow$ E)
  - test on voltage battery $V_{batt} \geq V_{max}$ ($\Leftrightarrow$ F).

From the number of test, we determined the number of possible combinations that we calculated using the following equation [17]:

$$X_c = 2^{n_{inp}} \tag{6.17}$$

where X number of possible combinations, n_{inp} number of inputs. We can obtain 64 combinations, but the number of possible combinations is reduced to 36. The logical equations are determined and give the control signals of the relays from each source: (Figs. 6.10, 6.11, 6.12)

$$\begin{aligned} T_{pv} &= \left(E.\overline{F} + \overline{E}.\overline{F}\right)\left(\overline{A}.D + \overline{A}.\overline{B}.\overline{C}.\overline{D} + \overline{A}.B.\overline{C}.\overline{D}.\overline{E}.F\right) \\ T_{wind} &= \left(E.\overline{F} + \overline{E}.\overline{F}\right)\left(\overline{B}.D + \overline{A}.\overline{B}.\overline{C}.\overline{D} + \overline{A}.B.\overline{C}.\overline{D}.\overline{E}.F\right) \\ T_{diesel} &= \left(E.\overline{F}.\overline{D}\right)\left(A.B + A.\overline{B}.\overline{C}. + \overline{A}.B.\overline{C}\right). \end{aligned} \tag{6.18}$$

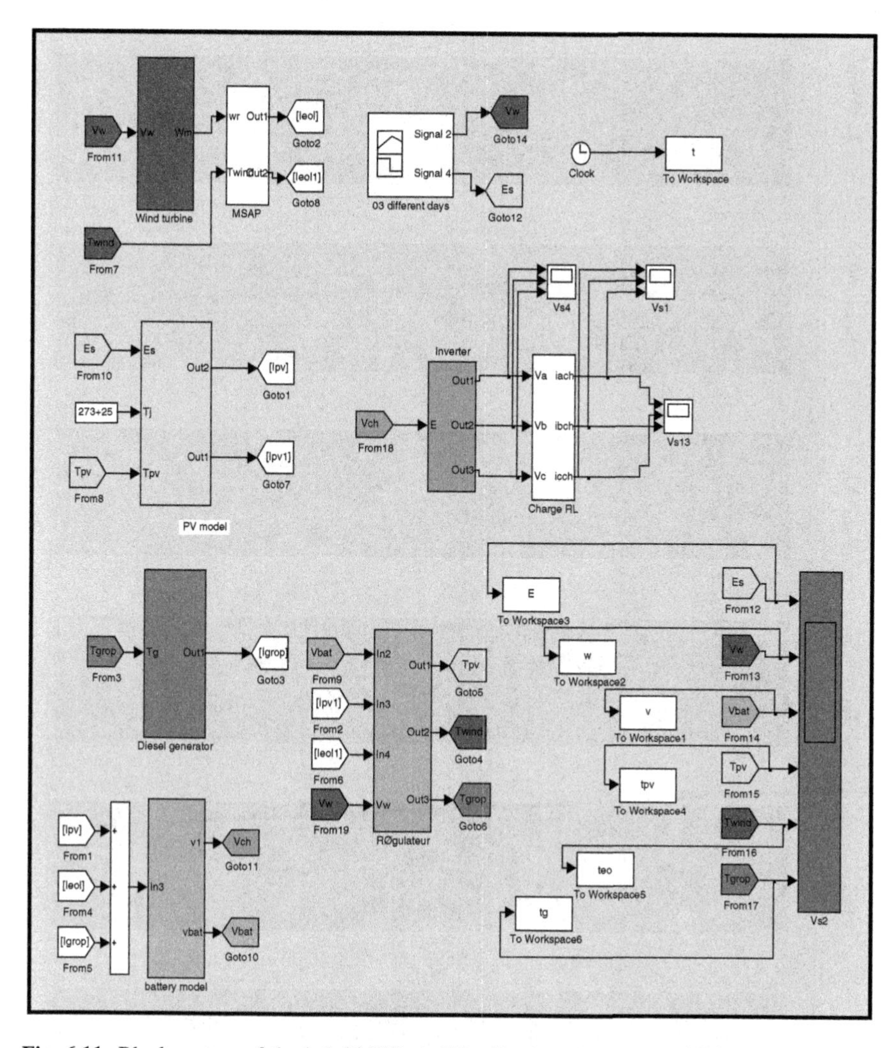

Fig. 6.11 Block system of the hybrid PV/wind/diesel generator system with battery storage

6.5 Hybrid Photovoltaic/Wind//Hydro System

These systems consist of microhydro, solar, and wind plants [18] (Fig. 6.13) and can be combined by a diesel generator backup.

The power control unit (PCU) is used to supervise and control the operations of PV/wind/hydro diesel hybrid power system. It coordinates when power should be generated by PV panels, wind turbine, and hydro turbine, and when it should be generated by diesel generator. The use of diesel generator is only when the demand cannot be sufficient by others energy.

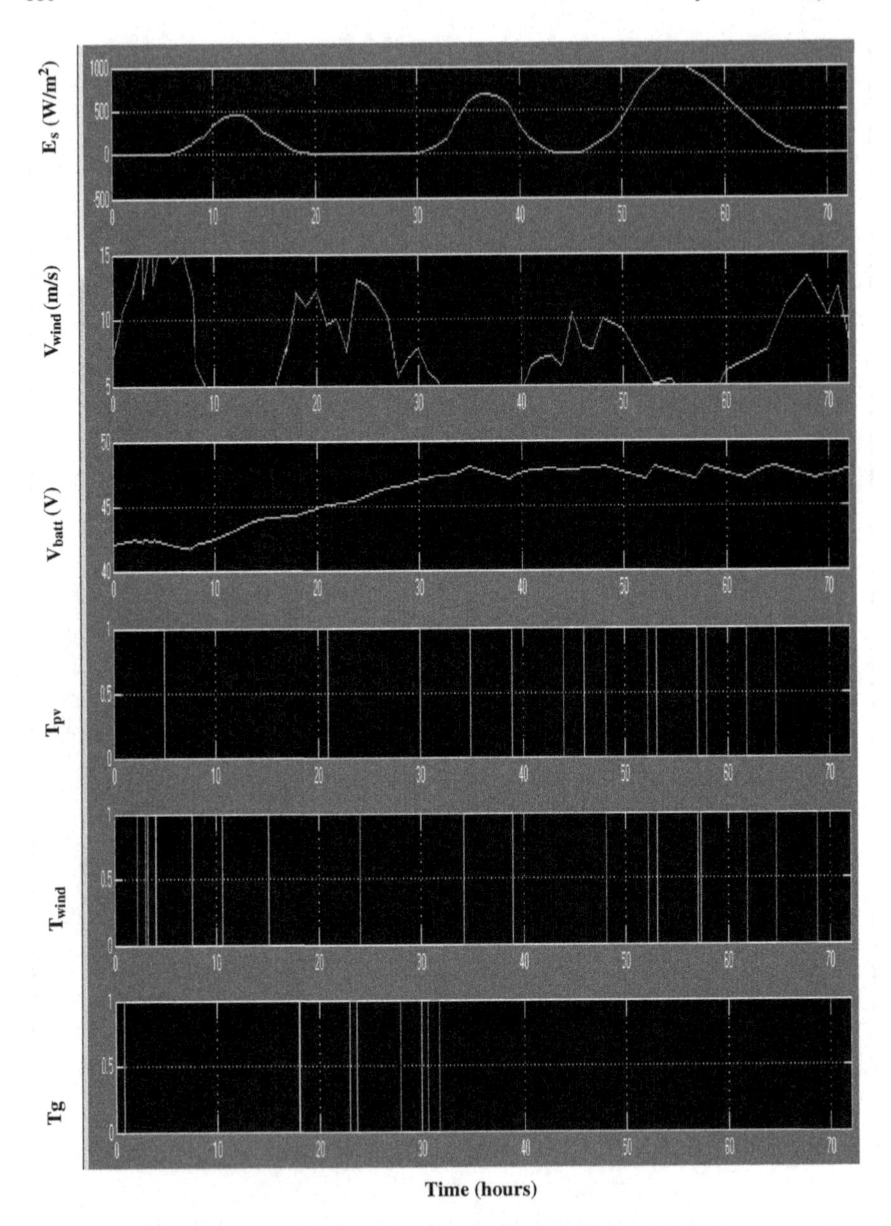

Fig. 6.12 Simulation results of hybrid wind/photovoltaic/diesel generator system with battery storage

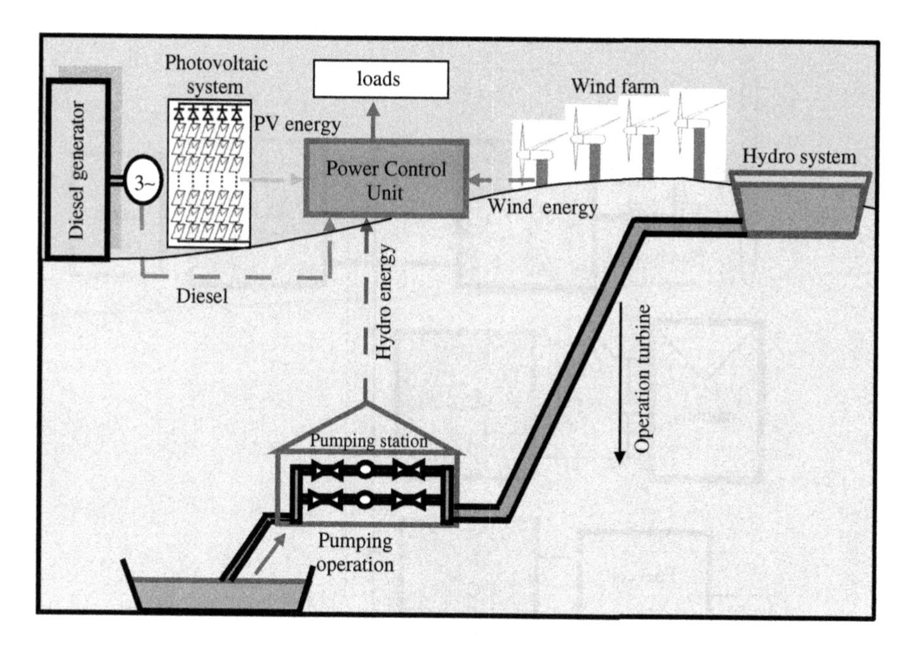

Fig. 6.13 Description of hybrid photovoltaic/wind//hydro system

6.6 Hybrid Photovoltaic/Wind/Fuel Cell System

The necessary changes in our energy supply system can be accomplished if we use a hybrid system with solar, wind energies, and fuel cell. Generally, the overall system comprises a wind subsystem with an AC/DC rectifier to connect the wind generator to the DC bus. It also consists of a PV subsystem connected to the DC bus via a filter and DC/DC converter. The excess energy is stored as electrolytic hydrogen through an electrolyzer, and we use a fuel cell to generate electricity during low insolation and low wind speed (Fig. 6.14).

In fact, when supply and demand do not coincide, we need a convenient way to both store and transport renewable energy. This is where hydrogen comes into play as a storage and transport medium. When excess electric energy from wind and solar energy is stored in hydrogen and then converted back to electricity, we have a solar–wind hydrogen energy cycle. Wind and solar, fuel cells, and electrolysis use excess electricity to split water into oxygen and hydrogen. When we need electricity the gases are fed into a fuel cell which converts the chemical energy of the hydrogen (and oxygen) into electricity, water, and heat.

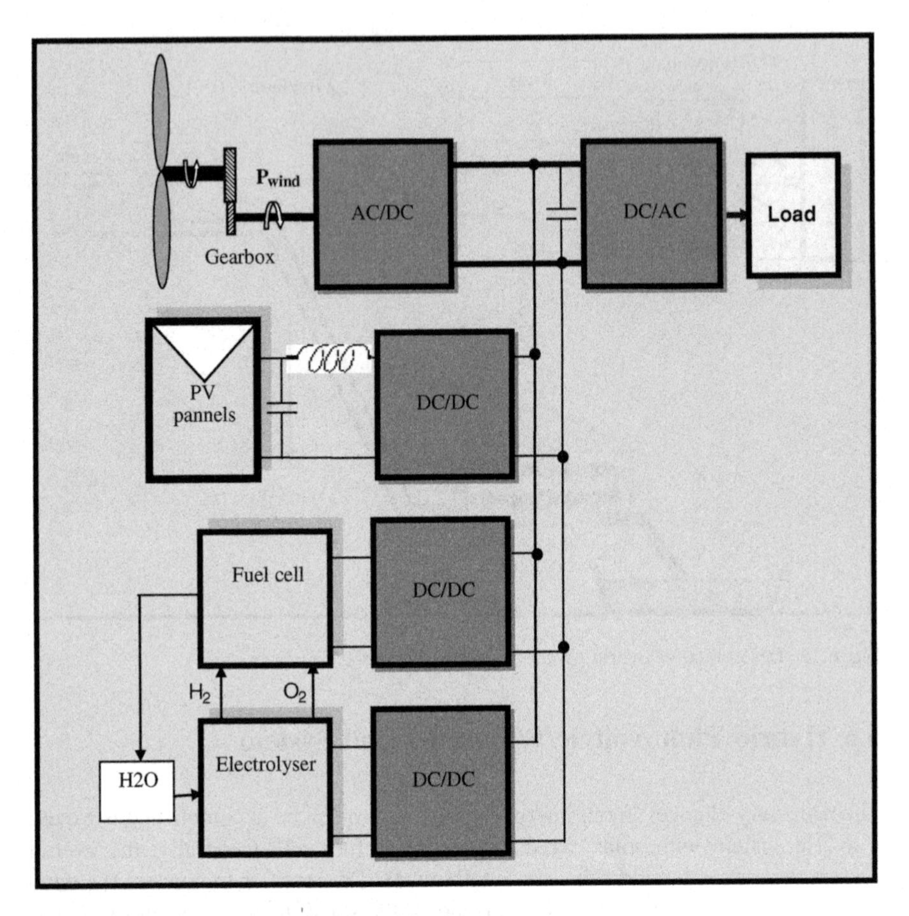

Fig. 6.14 Hybrid wind/PV/fuel cell configuration

6.7 Conclusion

This chapter has been devoted to hybrid wind systems. We have presented and described the different configurations and the different combinations of hybrid wind systems. Different synoptic schemes and simulation applications are also presented. We have presented a study of a hybrid wind/PV//diesel system with battery storage. The application is under Matlab/Simulink during three different profiles of insolation and wind speed (low, medium, and high conditions). This study seems interesting and can be applied to electrification or a pumping system for example.

References

1. Burton T, Sharpe D, Jenkins N, Bossany E (2001) Wind energy handbook. Wiley, London
2. Bianchi FD, De Battista H, Mantz RJ (2007) Wind turbine control system. Springer, London
3. Manwell JF, McGowan JG, Rogers AL (2002) Wind energy explained. Wiley, London
4. Ackermann T (2005) Wind power in power systems. Wiley, London
5. Saheb-Koussa D, Koussa M, Haddadi M, Belhamel M (2011) Hybrid options analysis for power systems for rural electrification in Algeria. Energy Procedia 6:750–758
6. Schubel PJ, Crossley RJ (2012) Wind turbine blade design. Energies 5:3425–3449. doi:10.3390/en5093425
7. Rekioua D, Rekioua T, Idjdarene K, Tounzi A, et al (2005) An approach for the modeling of an autonomous induction generator taking into account the saturation effect. Int J Emerging Electric Power Syst 4(1):1–25
8. Kishore A, Kumar GS (2006) Dynamic modelling and analysis of three phase self-excited induction generator using generalized state-space approach. In: Proceedings of international symposium on power electronics, electrical drives, automation and motion (SPEEDAM'06). IEEE, New York, pp 52–59
9. Malik NH, Al-Bahrani AH (1990) Influence of the terminal capacitor on the performance characteristics of a self excited induction generator. In: Transmission and Distribution, IEE Proceedings C 137(2):168–173
10. Nejmi A, Zidani Y, Naciri M et al (2002) Investigation on the self-excited induction generator provided with a hydraulic regulator. FIER, Tome II, Tétouane, Maroc, 8–10 May 2002, pp 494–499
11. Manwell JF, McGowan JG, Rogers AL (2010) Wind energy explained: theory, design and application. Willey, London
12. Nesba A, Ibtiouen R, Touhami O et al (2006) Dynamic performances of self-excited induction generator feeding different static loads. Serb J Electr Eng 3(1):63–76
13. Akon AF (2012) Measurement of axial induction factor for a model wind turbine. A thesis submitted to the College of Graduate Studies and Research in partial fulfillment of the requirements for the degree of Master of Science in the Department of Mechanical Engineering University of Saskatchewan, Aug 2012
14. Rahim Y. H. A (1993) Excitation of isolated three-phase induction generator by a single capacitor. Proc Inst Elect Eng Elect Power Appl 140(1):44–50
15. Elhafyani ML, Zouggar S, Benkaddour M, Zidani Y et al (2006) Permant and dynamic behaviours of self-excited induction generator in balanced mode. Maroccan Stat Phys Soc 7(1):49–53
16. Ibtiouen R, Benhaddadi M, Nesba A, Mekhtoub S, Touhami O et al (2002) Dynamic performances of a self-excited induction generator feeding different static loads. In: Proceedings of 15th international conference on electrical machine ICEM 2002, Brugge, 25–28 Aug 2002, pp 1–6
17. Wang YJ, Huang SY (2004) Analysis of a self-excited induction generator supplying unbalanced loads. In: Proceedings of international conference on power system technology (POWERCON'04). IEEE, New York 2004, pp 1457–1462
18. Poitiers F, Machmoum M, Zaim ME, Branchet T et al (2002) Transient performance of a self-excited induction generator under unbalanced conditions. In: Proceedings of 15th international conference on electrical machine ICEM 2002, Brugge. 25–28 Aug 2002, pp 1–6
19. Rekioua D, Matagne E (2012) Optimization of photovoltaic power systems: Modelization, simulation and control. Green Energy Technol 102

Chapter 7
Examples of Wind Systems

7.1 Examples of Wind Turbines

7.1.1 Wind Turbine of 600 W

In our Laboratory LTII (University of Bejaia—Algeria), we have installed a 600 W wind turbine (Fig. 7.1). Its technical specifications are given in Table 7.1.

Electrical specifications are summarized in Table 7.2 (Figs. 7.2, 7.3).

7.1.1.1 Identification of the PMSM Parameters

The machine used is a permanent magnet synchronous machine with rotor smooth, with an output of 600 W (Table 7.3).

To calculate the values of resistance and the stator inductance per phase, we use the following scheme (Fig. 7.4). In order to avoid the short-circuit of the source, we inserted a resistance R_s between the source and a phase stator.

Where R_s and L_s are, respectively, the stator phase resistance and inductance, R_s is the inserted resistance.

Three measurements of voltage and current were performed. We have

$$R_{\mathrm{ins}} = \frac{U}{I} \tag{7.1}$$

Thus, $R_{\mathrm{ins}} = 17.416\,\Omega$ and then, we calculate the stator resistance $R_s = 0.457\,\Omega$. To determinate the inductance, we have

$$Z = \frac{u}{i}$$

D. Rekioua, *Wind Power Electric Systems*, Green Energy and Technology,
DOI: 10.1007/978-1-4471-6425-8_7, © Springer-Verlag London 2014

Fig. 7.1 Example of a horizontal wind turbine of 600 W

Table 7.1 Technical specifications of 600 W wind turbine

Number of blades	02
Diameter	02 m
Material	Fiberglass and carbon fiber
Direction of rotation	Counterclockwise
Control systems	Electronic regulator

Table 7.2 Electrical specification 600 W wind turbine

Alternator	Three phase permanent magnet
Magnets	Ferrites
Nominal power	600 W
Nominal speed	1,000 rpm
Regulator	12 V 60 A

where

$$Z = \sqrt{(2R_s + R_{ins})^2 + 4L_s^2 \cdot \omega^2} \quad (7.2)$$

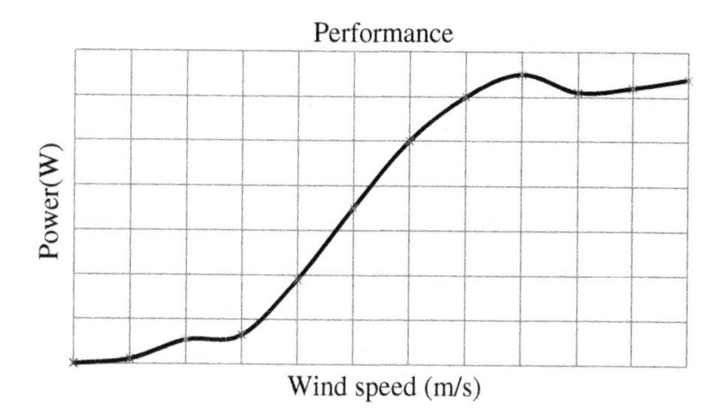

Fig. 7.2 Performance of 600 W wind turbine

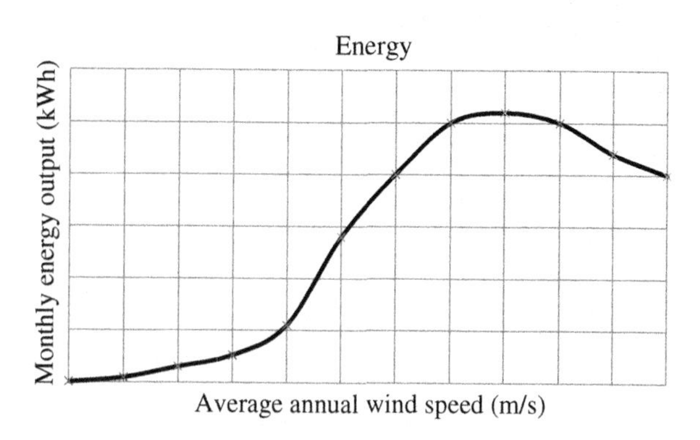

Fig. 7.3 Monthly energy output of 600 W wind turbine

Table 7.3 Electrical parameters of 600 W PMSM

Value	Parameter
P_N	600 W
R_s	0.457
L_s	0.029 H
τ_m	4.721 s
ϕ_f	0.13 Wb
P	17
J	0.1 kg m^2
f	0.06 N m s/rad

Thus,

$$L_s = \sqrt{\frac{(u/i)^2 - (2R_s + R_{ins})^2}{4 \cdot \omega^2}} \tag{7.3}$$

We obtain $L_s = 0.029\,\mathrm{H}$

A wind speed of 17 m/s, the electrical speed $\omega_e = 196.349$ rad/s; thus, $P = 17$.

The excitation flux is $\Phi_f = 0.13\,\mathrm{Wb}$, the rotor inertia is $J = 0.1$ kg m^2, and the viscous coefficient is about $f = 0.06$ N m s/rad (Fig. 7.5).

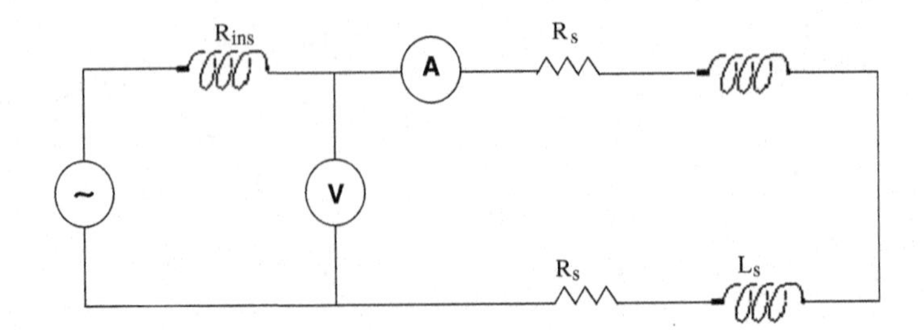

Fig. 7.4 Diagram of two-phase stator fed by a DC or AC source

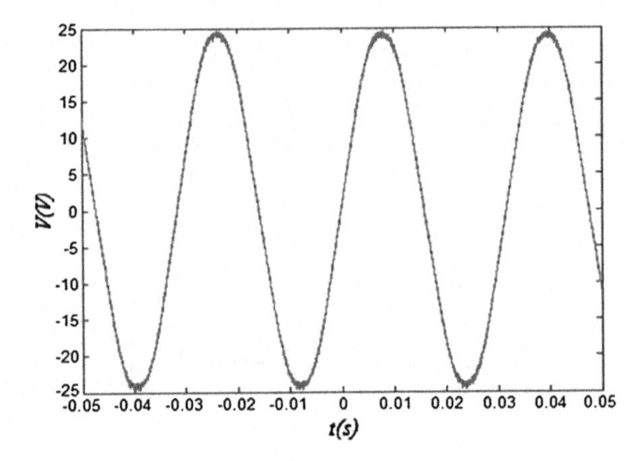

Fig. 7.5 Voltage at wind speed of $V_{wind} = 17$ m/s

7.1.1.2 Simulations Results

We make an application of the WECS of 600 W under MATLAB/SIMULINK. We choose a wind speed profile (Fig. 7.6) (Figs. 7.7, 7.8, 7.9, 7.10 and 7.11).

We remark that the electrical power follows the wind speed profile with a maximum value of power which corresponds to a wind speed of 13 m/s.

7.1.2 Wind Turbine of 1 kW

The turbine comprises a permanent magnet brushless alternator, which combined with Whisper's high efficiency composite airfoil blade design, delivers 900 W peak power at 28 mph (12.5 m/s). It is designed to operate with medium to high wind speed averages of 12 mph and greater (see Tables 7.4 and 7.5) (Fig. 7.12).

Its applications cover stand-alone or hybrid systems, telecommunication, remote home and small applications (Figs. 7.13, 7.14 and 7.15).

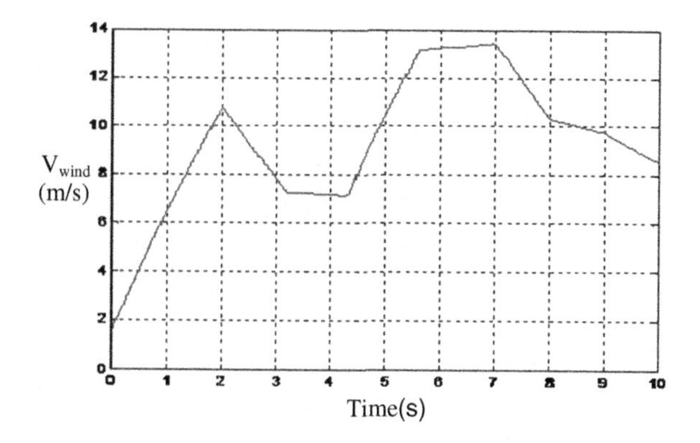

Fig. 7.6 Voltage at wind speed of $V_{\text{wind}} = 17$ m/s

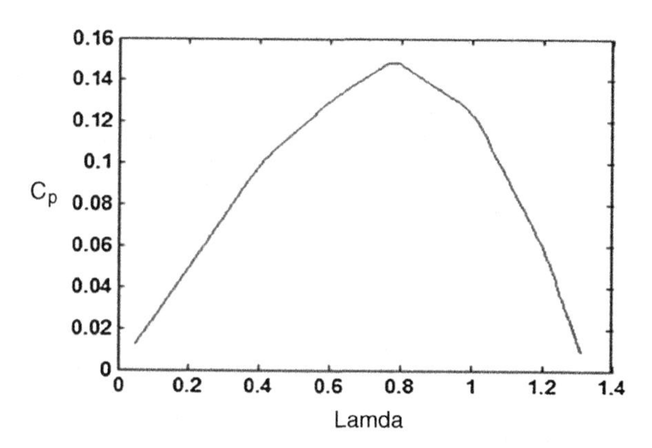

Fig. 7.7 Power factor

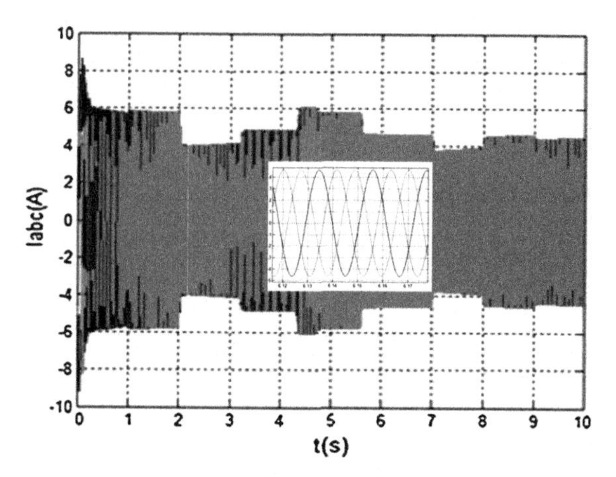

Fig. 7.8 Current waveforms

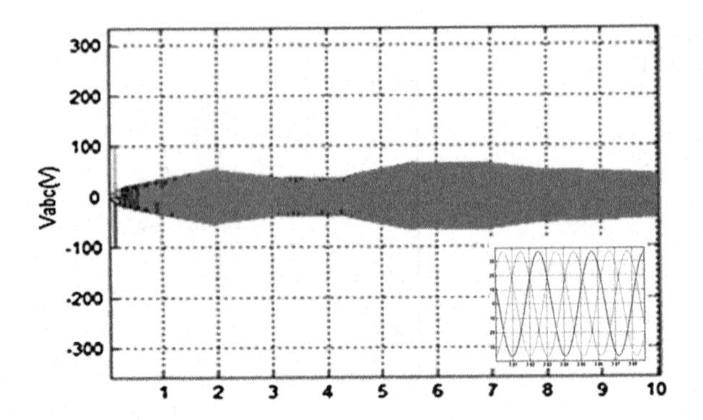

Fig. 7.9 Voltage waveforms

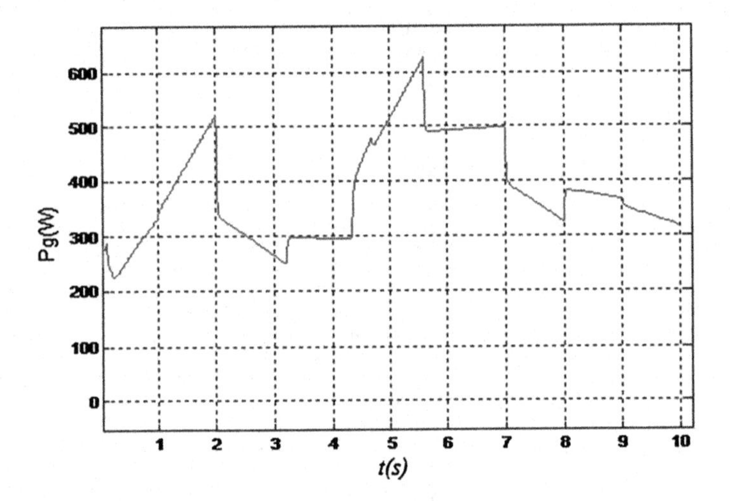

Fig. 7.10 Power waveform

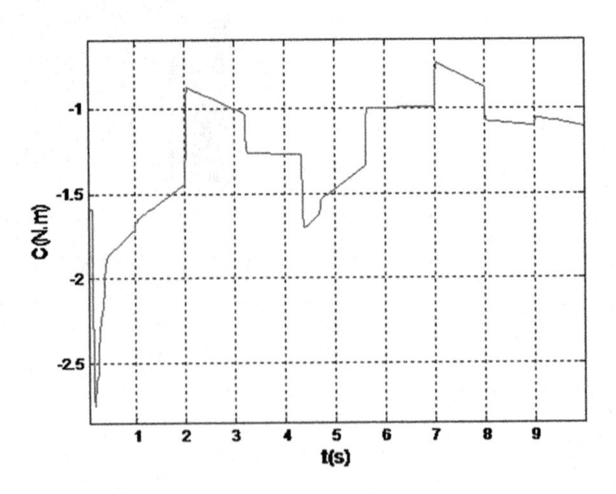

Fig. 7.11 Electromagnetic torque

Table 7.4 Technical specifications of wind turbine of 900 W

Rotor diameter	2.1 m
Voltage	12 V, 24 V DC
Start up wind speed	3.4 m/s
Survival wind speed	55 m/s
Over speed protection	Side furling
Rated power	900 W at 12.5 m/s
Turbine controller	Whisper controller
Kilowatt hours per month	100 kW h/month at 5.4 m/s

Table 7.5 PMSM electrical parameters of 900 W

Values	Parameters
P_N	900 W
R_s	0.49
L_s	0.0016 H
τ_m	4.721 s
ϕ_f	0.148 Wb
P	5
J	1.6
f	0.0001

Fig. 7.12 Example of a horizontal wind turbine of 900 W

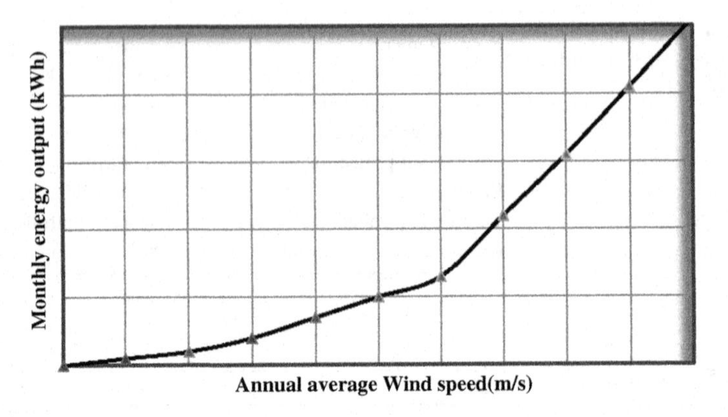

Fig. 7.13 Monthly energy output

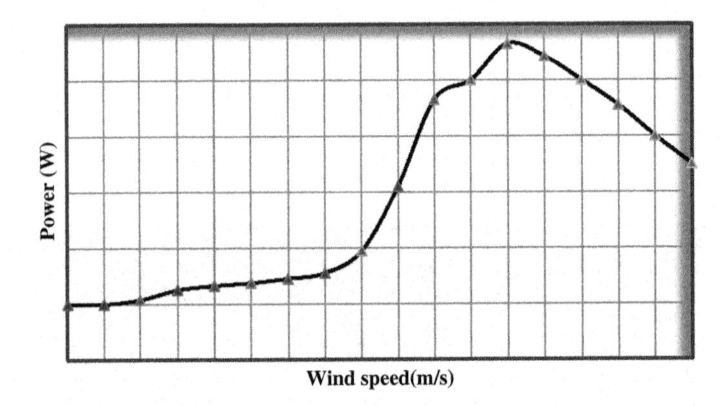

Fig. 7.14 Output power

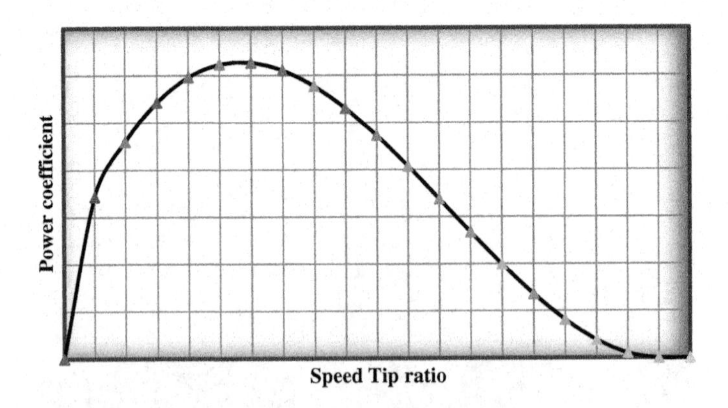

Fig. 7.15 Power coefficient of wind power 900 W

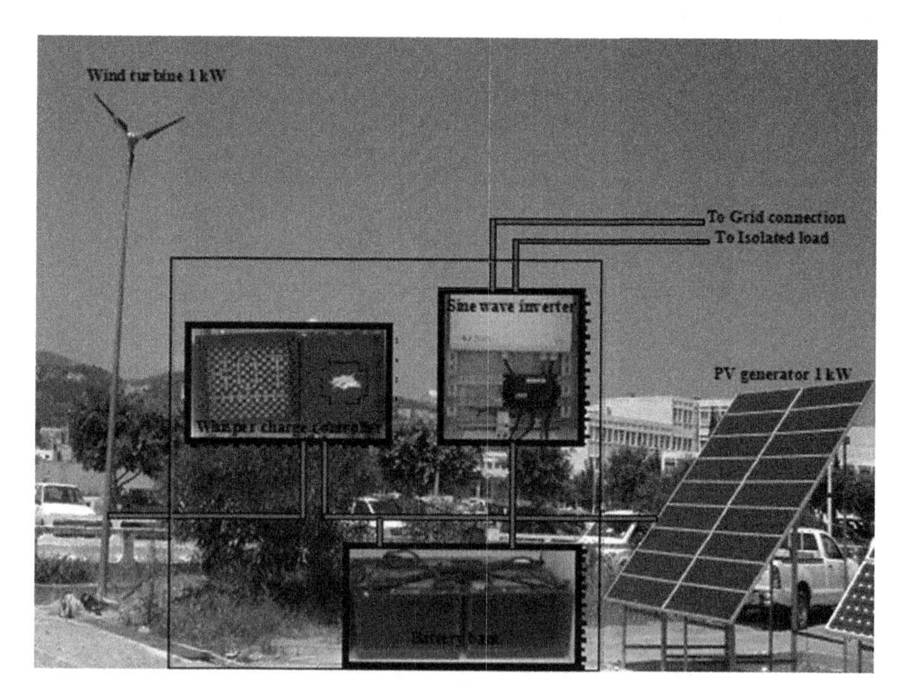

Fig. 7.16 Installed hybrid wind–photovoltaic system

Fig. 7.17 Battery bank

The overall installed system in our laboratory comprises a wind turbine; a Whisper controller (Fig. 7.16), which offers greater reliability and superior control for battery charging; a battery bank (Fig. 7.17); fuses; and electrical protection cards (Fig. 7.18).

Fig. 7.18 Whisper controller with cards protection and fuses

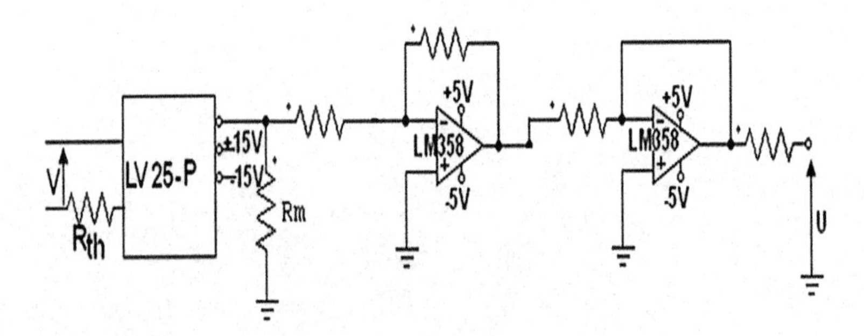

Fig. 7.19 Voltage sensor

For voltage sensors (Fig. 7.19), we use Hall effect sensor LV25P equivalent to a transformer and which is composed of a primary coil and a secondary coil.

For current sensor, its principle is the same as that of the voltage one, measuring current creates an output voltage proportional. It differs from the voltage sensor by the number of input pins and does not require resistor to limit the input current as the previous sensor (Fig. 7.20).

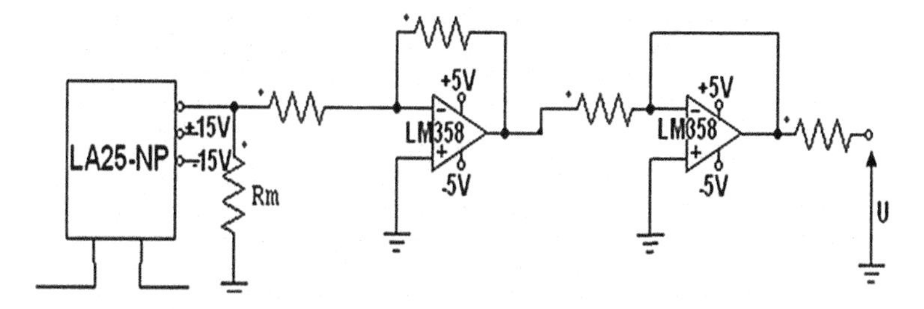

Fig. 7.20 Current sensor

Fig. 7.21 Wind turbine tests

7.1.2.1 PMSM Parameters Identification

Some tests are important to confirm that the wind generator was not damaged in shipment and is ready to install on the tower (Figs. 7.21, 7.22, 7.23, 7.24, 7.25 and 7.26).

- Ground test
- Open-circuit test
- Short-circuit test
- Phase to phase test.

Fig. 7.22 PMSM of the wind turbine

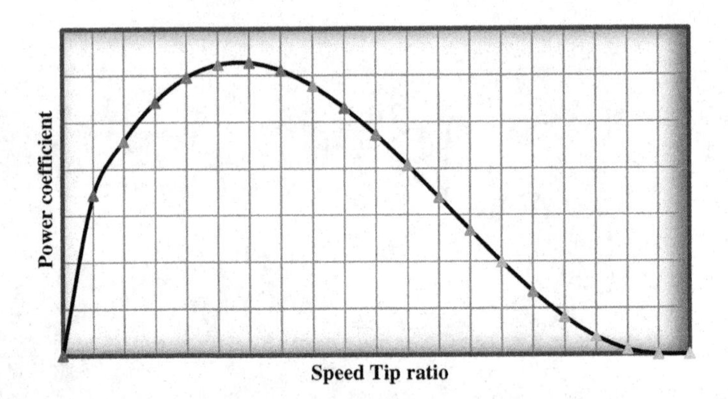

Fig. 7.23 Power coefficient

7.1.2.2 Practical Results

An anemometer (Fig. 7.27) with wind sensor, temperature sensor, and humidity sensor is arranged on the wind turbine with an indoor unit display (Fig. 7.28).

We present some results of signal measured at different frequencies (Figs. 7.29, 7.30, 7.31, 7.32 and 7.33).

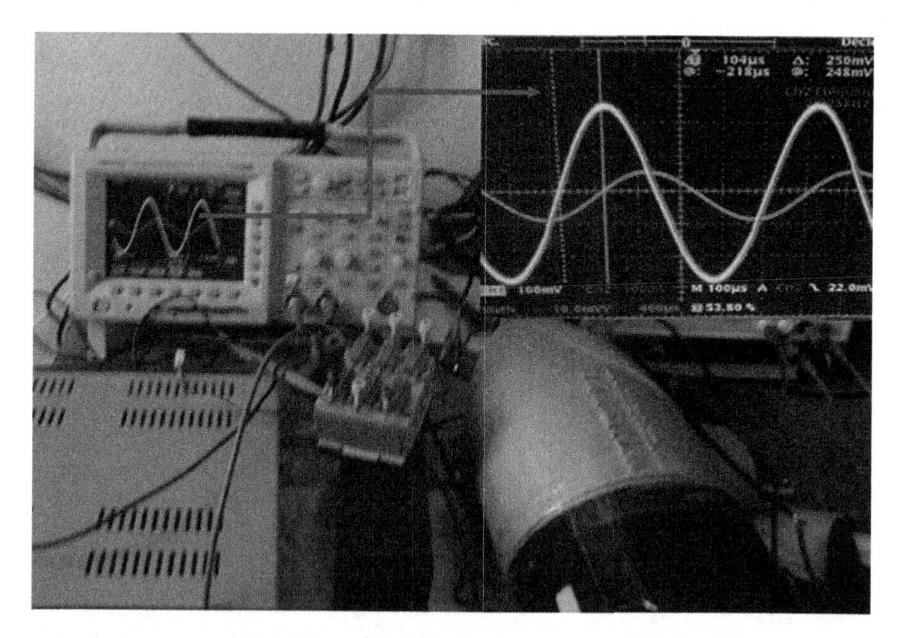

Fig. 7.24 Voltage and current waveforms

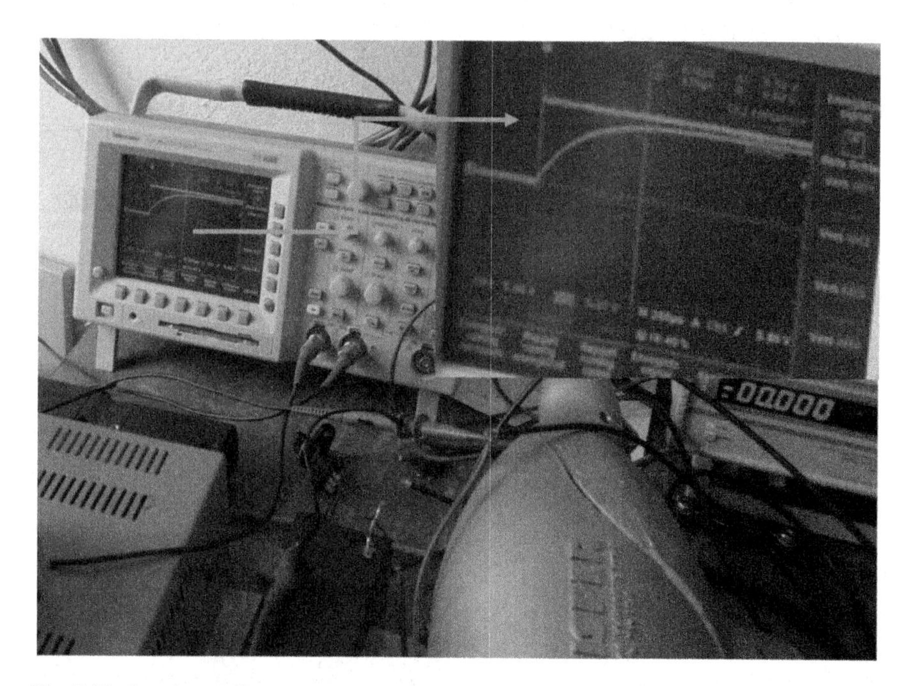

Fig. 7.25 Speed waveform

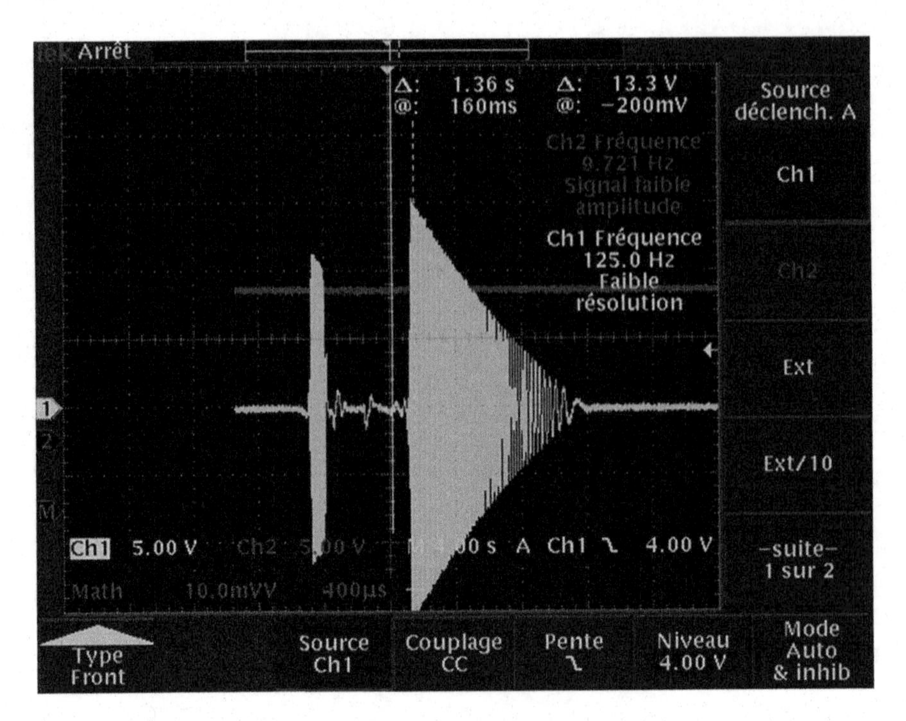

Fig. 7.26 Braking test machine

Fig. 7.27 Anemometer

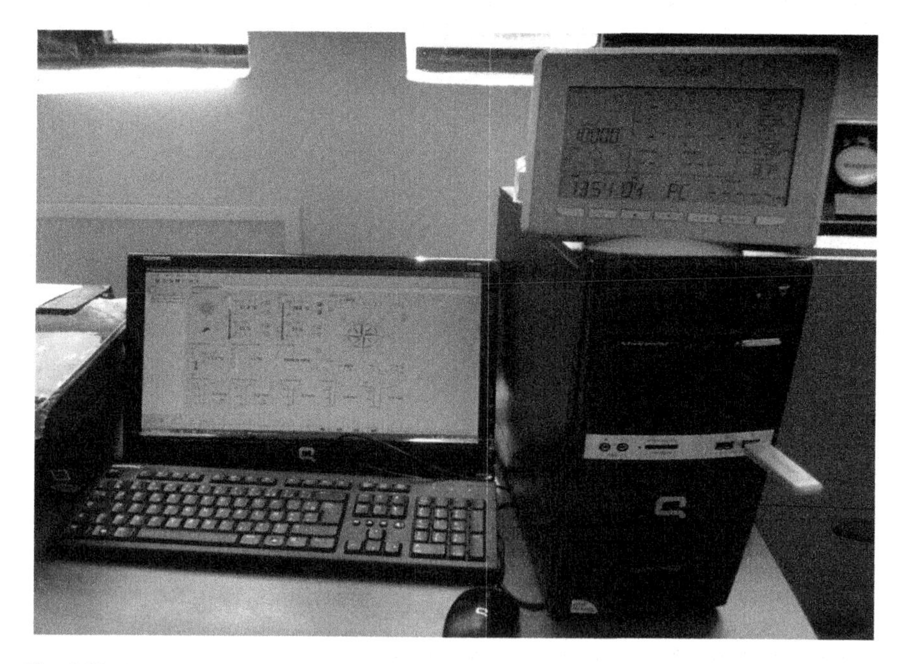

Fig. 7.28 Indoor unit display of the anemometer

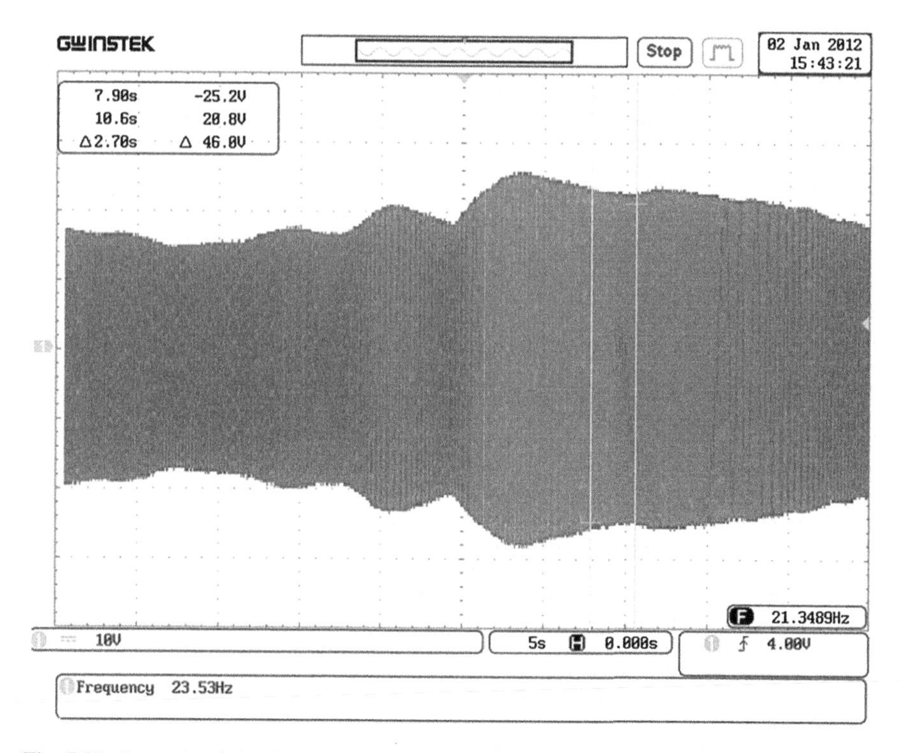

Fig. 7.29 Example of signal at $f = 21.348$ Hz

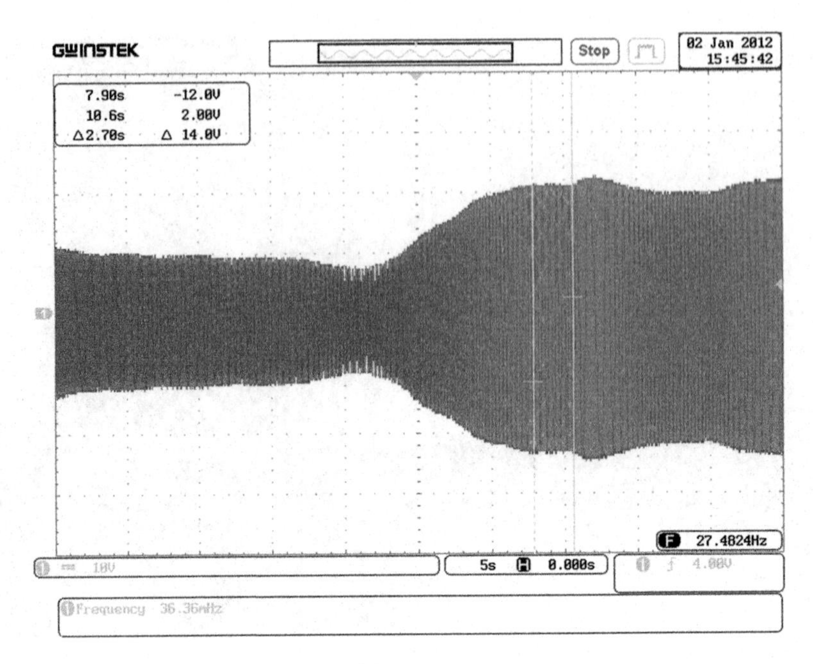

Fig. 7.30 Example of signal at $f = 27.482$ Hz

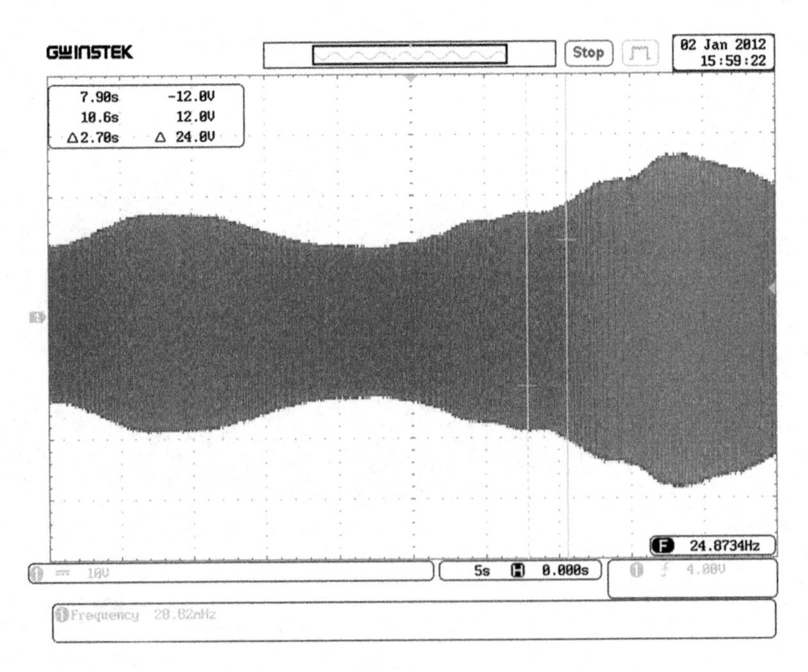

Fig. 7.31 Example of signal at $f = 24.873$ Hz

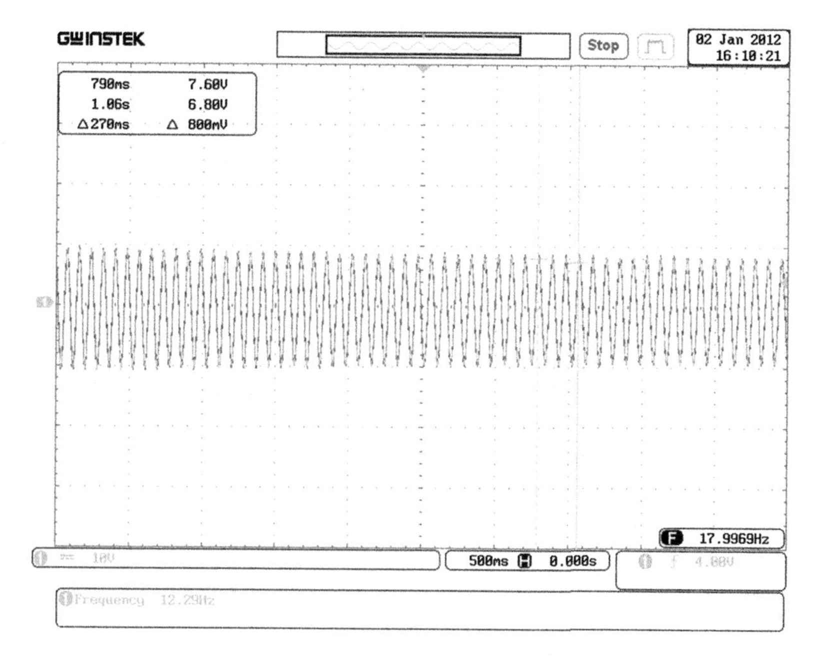

Fig. 7.32 Example of signal at $f = 17.9969$ Hz

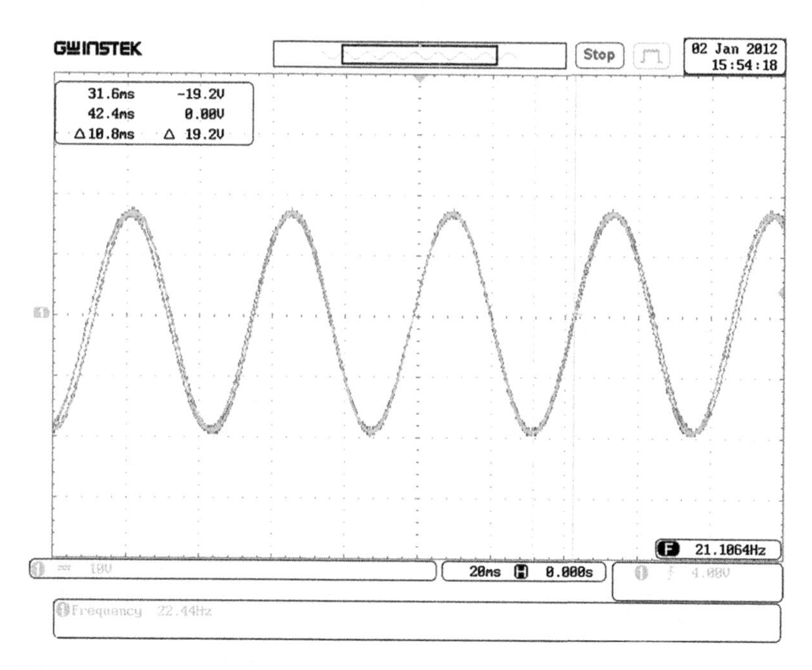

Fig. 7.33 Example of signal at $f = 21.106$ Hz

7.2 Conclusion

In this chapter, we have presented two wind examples with simulation and experimental tests. These examples can help students of graduation and post-graduations in the fields of electrical engineering quickly understand the identification of wind turbine and make easily the simulation of the studied system.

Printed in the United States
By Bookmasters